3D MOTION GRAPHICS
FOR 2D ARTISTS

3D MOTION GRAPHICS FOR 2D ARTISTS

Conquering the Third Dimension

BILL BYRNE

ELSEVIER

AMSTERDAM • BOSTON • HEIDELBERG • LONDON • NEW YORK • OXFORD
PARIS • SAN DIEGO • SAN FRANCISCO • SINGAPORE • SYDNEY • TOKYO
Focal Press is an imprint of Elsevier

Focal Press

Focal Press is an imprint of Elsevier
225 Wyman Street, Waltham, MA 02451, USA
The Boulevard, Langford Lane, Kidlington, Oxford, OX5 1GB, UK

Notices

Knowledge and best practice in this field are constantly changing. As new research and experience broaden our understanding, changes in research methods, professional practices, or medical treatment may become necessary.

Practitioners and researchers must always rely on their own experience and knowledge in evaluating and using any information, methods, compounds, or experiments described herein. In using such information or methods they should be mindful of their own safety and the safety of others, including parties for whom they have a professional responsibility.

To the fullest extent of the law, neither the Publisher nor the authors, contributors, or editors, assume any liability for any injury and/or damage to persons or property as a matter of products liability, negligence or otherwise, or from any use or operation of any methods, products, instructions, or ideas contained in the material herein.

Library of Congress Cataloging-in-Publication Data
Byrne, Bill, 1976-
  3D motion graphics for 2D artists : conquering the 3rd dimension / Bill Byrne.
     p. cm.
  ISBN 978-0-240-81533-6 (pbk.)
  1. Three-dimensional imaging. 2. Computer animation. 3. Computer graphics. I. Title.
  T385.B935 2012
  006.6--dc23
                        2011038126

British Library Cataloguing-in-Publication Data
A catalogue record for this book is available from the British Library.

ISBN: 978-0-240-81533-6

For information on all Focal Press publications
visit our website at *www.elsevierdirect.com*

11 12 13 14 15   5 4 3 2 1

Printed in the United States of America

Typeset by: diacriTech, Chennai, India

Working together to grow
libraries in developing countries

www.elsevier.com | www.bookaid.org | www.sabre.org

ELSEVIER   BOOK AID
            International   Sabre Foundation

Dedication

To my wife

Suzanne, for her love and support that makes everything I do possible. Thank you for understanding during my lapse of progress on weekend projects.

To my daughter

Elinor, in spite of her best efforts to hinder the writing of this book, through keyboard grabbing and distracting me by being absolutely adorable, it did get finished. Da-da loves you.

CONTENTS

ACKNOWLEDGMENTS

My parents, Tom and Marie Byrne.

My closest friends, Larry Caldwell, Jonah Goldstein, and Bryan Wetzel, for their advice and council.

The Art Institute of Austin's Dean of Academic Affairs, Regina Verdin, former Dean, Carol Kelley, and President, Monica Jeffs.

My colleagues and good friends, Andrea Alexander, Peggy Blum, William Gatz, Elizabeth Jenkins, Conrad Rathmann, Mark Sarisky, and Barry Underhill.

My friends and interview subjects Mayet Bell, Luc Dimick, Dax Norman, and Erica Hornung.

The students in the Media Arts and Animation, Game Art and Design, Visual Effects and Motion Graphics, and Digital Film and Video Production departments at the Art Institute of Austin, for whom a book like this was sorely needed.

Focal Press, Dennis McGonagle, Carlin Reagan, and Laura Aberle, who made all of this possible.

INTRODUCTION

This book is designed to solve a very specific problem—flatness. Two-dimensional motion graphics designers and animators routinely hit a brick wall. You have these flat objects occupying a manufactured world, and the illusion of depth is just not enough—you want it to turn in *space*. More so, you have a client who is demanding that object turn in space, and you need to meet that request (or, let's face it, someone else will).

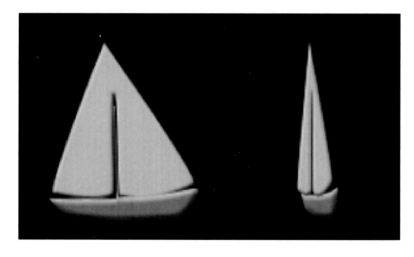

Figure 1.1 2D object turning in 3D space.

This usually leads folks to look for plug-ins for After Effects, or a three-dimensional animation application. Then a new roadblock appears (or several)—how do I use these things? First, 3D is *a lot* more difficult than 2D. Second, beginner resources for 3D software start you from the very, very beginning, and you may not have the time to spend spinning cubes.

This book will explore several options for 2D motion graphics artists. First, we'll take advantage of the new and not-so-new 3D tools in Adobe software, which are quite useful and may solve a large percentage of your concerns. Second, this book will address existing tools that can be added to After Effects, like third-party plug-ins that add 3D functionality to AE. Third, we'll dip into 3D software including Cinema 4D, ZBrush, 3dsMax, and others, starting you in the direction of developing your 3D skill set.

TIP

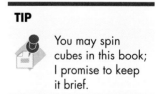

You may spin cubes in this book; I promise to keep it brief.

How to Use This Book

If you are familiar with either of my two previous textbooks, *The Visual Effects Arsenal* (Focal Press 2009) and *Creative Motion Graphics Titling* (coauthored with Yael Braha; Focal Press 2010), then you won't find what I am about to say all that surprising. My teaching motivates my textbook writing. I am as fascinated by *how* people learn as I am by *what* they learn.

Students and professionals want books that address specific problems and solve them. They are not nearly as motivated by the guts of a software package as they are by what it can do for them. This book is designed to take people who know 2D motion graphics and animation and get them on the right path to learning 3D. This book won't teach you everything you need to know about 3D motion graphics—learning everything about 3D animation is a lifelong pursuit.

The aim of this text is to get you started in 3D by introducing you to available options that can be added with relative ease to an existing 2D animation skill set. In many ways, this book is designed to be 3D animation from the very, very beginning. This book does assume some foreknowledge of standard 2D tools (such as Adobe After Effects, Adobe Photoshop, and Adobe Illustrator). 3D packages will be introduced like 3dsMax, Cinema 4D, ZBrush; and After Effects plug-ins like Trapcode Particular, Trapcode 3D Stroke, and more.

Even though you will be introduced to numerous tools that have infinite applications, techniques will be the focus of this book. What's the point of learning an incredibly complex tool if you don't get to see what it does?

THE TECHNIQUES OF 3D

Creating the Illusion of Depth

Figure 2.1 *The Grid Is Broken* digital painting by Bill Byrne.

The essential point of creating 3D art and animation is to create a believable world. That world can be completely false, but the audience needs to buy into your reality to some degree or it will be meaningless to them. Even the strangest realities will need to have some signposts for the audience to be able to get some sense of it. In any case, the 3D artist must build a world, a world that is also a stage.

What's the difference? A world exists, and does things with or without the viewer. It pays the viewer little mind. A stage contains a representation of something that resembles the world for the benefit of the viewer. It's a show. Art is always a presentation for the viewer. Flat art, such as drawings, photographs, films, and paintings have always had a limitation in that they pretend to contain true depth but are actually flat illusions of depth.

Why does this matter to you? Well, 3D animation will take the real world and recreate it within a flat space. It matters to you because in this day and age, when that "wow" factor is few and far between, what better way to engage viewers than inviting them into a completely imagined world that they can relate to and that is a complete work of your own fantasy and design?

Three Dimensions

Objects in our world have three dimensions: height (represented by X space in Figure 2.2), width (Y), and depth (Z). Nonsculptural, flat art represents height and width easily because flat art is two-dimensional. Now you may be thinking, I have seen three dimensions in flat art, but that actually is not true. You've seen the *illusion* of depth. In fact, there are a series of techniques that are employed to make something appear to have depth that in fact does not.

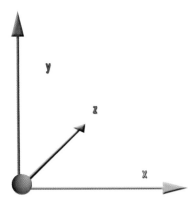

Figure 2.2 Three dimensions.

Stacking or Layering

Figure 2.3 The image on the left shows various objects unstacked. The image on the right shows the same objects stacked. Notice how by simple placement of the objects on top of each other the image starts to take on a certain amount of depth.

The first of these techniques is *stacking* or *layering*. What this simply means is placing objects on top of each other. Objects that are closer to the viewer should be on top of objects that are farther from the viewer.

Stacked or layered objects overlap others and this establishes distance quickly and effectively.

Top and Bottom Placement

The vertical order in which objects are arranged in a frame influences the perception a viewer has of how close or far an object is from them.

Figure 2.4 In a landscape, the higher an object is placed in a frame, the farther it appears to be from the viewer.

However this rule varies based on the content of your frame. For example, in Figure 2.4, an exterior space or landscape's depth is determined by placing objects closer to the top of the frame if they are farther away from the viewer. In an interior space, the objects at the top of the frame appear closer to the viewer; see Figure 2.5.

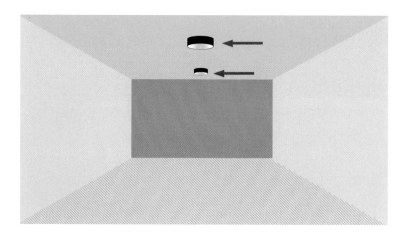

Figure 2.5 In an interior space, the ceiling reverses the rule we discussed for exteriors— the top of the frame is closer to the viewer. Think of a ceiling and floor as a landscape with a ground area at the top and bottom of the frame.

Color

The use of color can also create depth in an image. Objects closest to the viewer would be the most saturated, with the brighter colors. Objects farther away will be darker or less saturated. In landscapes such as Figure 2.6 you will see that the mountains in the distance look as if they have a blue-gray tint.

Figure 2.6 Color saturation levels also create depth in an image. Notice in this landscape photo that the far-off mountains have a blue-gray hue.

Perspective

Perspective is quite possibly the most effective technique for creating depth in a flat image. Lines that occur naturally within an image moving toward a *vanishing point* create perspective. The vanishing point is where the lines creating perspective converge, gesturing a continuation that is no longer perceived by the human eye. Perspective lines can be used as a guide to keeping objects following a diminishing scale. Let's say we take two objects that are believed to be the same size by the viewer. The object that is larger in the frame will be perceived as closer to the viewer.

Figure 2.7 Photographic representation of depth and perspective.

Multipoint Perspectives

Figure 2.8 The blue point is the single vanishing point.

One-point perspective is employed for classic road-goes-on-forever, railroad-tracks-to-infinity-type images, because it employs a single vanishing point. Every line in the image meant to indicate perspective moves to that vanishing point.

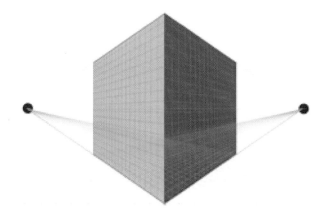

Figure 2.9 A two-point perspective uses two vanishing points to demonstrate the depth of an object that has two separate planes.

Two-point perspective would be used to take an object in perspective and rotate it on the horizontal axis. Think of a photograph of the corner of a building. The two sides of the building are following two separate planes, and therefore the perspective should be adjusted accordingly.

To demonstrate a top-down bird's-eye view of an object, *three-point perspective* can be employed. This time there would be three vanishing points employed to present the object (see Figure 2.10).

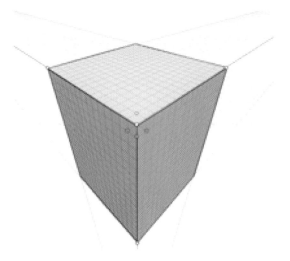

Figure 2.10 A three-point perspective uses three vanishing points to show a top-down view of an object.

Depth of Field

By emulating the behavior of the camera, we can also create a sense of depth. The camera's lens works like a human eye; the objects in focus appear to be closer than objects that are not sharp. However, cameras also allow us to flatten space by leveling the amount of change of focus within an image or depth of field.

Defining the Difference between 2D and 3D Computer Graphics

The techniques just discussed work when applied to 2D art in creation of the *illusion* of depth. We will discuss several techniques for creating the illusion of depth in 2D graphics, but what about when we need to get past that? 3D graphics software is designed to create virtual worlds like our own real one, within our computers, to handle all of this illusion stuff for us. So rather than painstakingly emulate the behavior of light, the software tools we will discuss will be able to apply algorithms to this for us.

I know what you might be thinking: If I can take a photograph of the real world, why is that considered 2D when a scene built in Maya is considered 3D? The answer is essentially the following: A photograph on a computer monitor is just flat pixels assigned to specific positions. A 3D scene uses wireframe structures that can be viewed from any angle, by moving a virtual camera. As discussed earlier, it is a virtualization of the real world on your computer that can mimic our real world as closely as the artists involved are willing to make it.

Questions about 3D You Were Afraid to Ask

We 3D artists can be an elite bunch with little time for questions from noobs. This section is designed for you beginners, with some of the best answers I could come up with for very common questions.

What Is a 3D Model?

Models are mathematical shapes formed into the surface of 3D objects. Vector objects are created by the positions of points in a 3D space, and then a raster graphic is assigned that geometric object to define its appearance. So the vector wireframe can be moved around within the space where its size, angle, and position can be adjusted at any point in time. Models are either built by hand or from a scanned object. In this book we will address the process of making models by hand, as in many cases it is just not possible to use a 3D scanner on an object (like a building).

What Is Texture Mapping?

A model is a wireframe with a surface. That surface can be a flat color, an image that was drawn or painted, or even a photograph. So texture mapping is the process of applying a raster image surface to a 3D model. Several different maps can be applied to the same object to create what is known as a *multitexture*. These would typically include a *UV Map*. UV mapping is the process of creating maps where specific points in a 2D texture are plotted to specific locations on a 3D model.

What Is Rendering?

Rendering is the process of creating a 2D image, be it moving or still, of a 3D model or scene consisting of a number of models.

Why Is Poly Count an Issue?

Every 3D object is made of polygons. A sphere can be infinitely subdivided into a larger number of polygons. So, the more polygons used, the more complex the surface of an object can be. However, the more complex the objects are, the longer it would take to render. Video game designers typically go for lower poly counts because they want the rendering to happen more quickly (since rendering in a game is real-time). Very high poly counts mean the more photo-real the 3D object will be.

What Is the Difference between 2.5D and 3D?

2.5D refers to techniques that software will use in order to make a hybrid that lives somewhere between 2D and 3D. For example, Adobe After Effects has a 3D switch that allows users to turn a 2D object in 3D space. That object is still completely flat—it is just turning in 3D space.

THE TOOLS OF 3D

We are assuming that you are a knowledgeable 2D motion graphics artist. We figure you are someone who is looking to branch out your skills by adding new software skills. This book does assume a background with the Adobe CS suite, mainly Adobe Photoshop, Adobe Illustrator, and Adobe After Effects. The first few chapters will address techniques for these three packages that are either (1) newly added to the software or (2) classic techniques that users may just not be aware of. Photoshop in its last three versions has added significant and powerful 3D tools. After Effects has supported 2.5D for several versions. With version CS4 it also supports Photoshop 3D files. Additionally there's a long history of add-on tools or plug-ins that add 3D support.

However, this book also addresses how you would dip into other software to get tools like After Effects, which the Adobe suite just doesn't come with. So we will look at 3D hosts, modeling tools, digital sculpture tools, and other options for working some 3D into your motion graphics. The list and advice that follow are in no way intended to be comprehensive or set in stone. At every opportunity I try out a tool or software package that may improve my workflow and I encourage you to do the same. You know your own needs better than anyone else, and part of the theme of this text is managing your expansion into 3D on an as-needed basis.

The Essentials

Image Editing: Photoshop

Adobe Photoshop is an absolutely essential software package for 3D or 2D, or for any person who wants to make an image on a computer. There was a time when people used it primarily and only to edit photos but that's a ridiculous notion today. 3D artists have been using it forever, and now with some new tools there are more reasons to own it than ever.

I remember being completely blown away when Photoshop CS3 allowed you to import Google Sketchup models. CS3 also

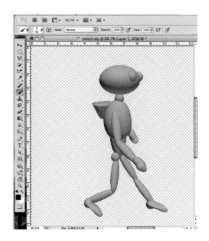

Figure 3.1 Editing a 3D model in Photoshop.

introduced a keyframe-based animation timeline, which is much like After Effects'. CS4 completely blew the lid off of that by allowing users to paint directly on OBJ models. Now, CS5 has its own 3D extrusion tool, *Repousse*. It's starting to feel as if Adobe thinks that someday this could be some kind of meta-program.

Additionally, editing raster files for UV maps has been a basic role of Photoshop in a 3D workflow for a long time. Generating and painting textures as well as working with photographic images are also among this powerhouse program's strong suits.

Motion Graphics: After Effects

Adobe After Effects (AE) was once pitched as Photoshop on a timeline. This ability of After Effects has made it so useful that it has become ubiquitous in the motion graphics, visual effects, and animation industries. After Effects works seamlessly with the Adobe suite, most notably Photoshop, Illustrator, Premiere Pro, and Flash (to some extent).

It has a raster-based engine, but it will take assets that are raster, vector, video, or audio. 3D models can be imported, but they would have to go to Photoshop first before they could be used in AE. AE is by nature a 2D/2.5D keyframe animator with limited true 3D capabilities. In addition to that it can be used for compositing and color correction.

One of AE's best features in its expandable toolset is allowing users to purchase third-party plug-ins. Third-party plug-ins are available from many different publishers for varying costs. We'll get a little deeper into that end later in this chapter.

AE's main competitor is Apple's Motion. However, after being in publication for about seven years, Motion has failed to unseat After Effects' position at the top of the heap.

After Effects 3D Plug-Ins

After Effects' customizable plug-in system has led to the development of lots of great add-on tools. Following are some of the best tools for adding 3D functionality to After Effects.

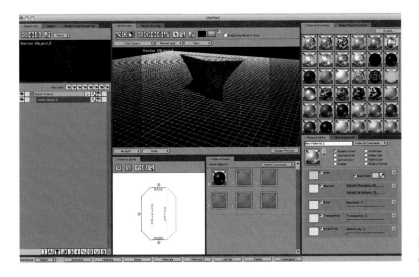

Figure 3.2 Zaxwerks Pro Animator.

Zaxwerks Pro Animator and Plug-In Product Line

Zaxwerks' 3D plug-ins for After Effects have been adopted by the industry, with good reason—they add a large number of 3D capabilities with very little learning curve. The top of their product line is the Pro Animator plug-in/stand alone application.

Pro Animator came from the features they have long offered in their classic 3D Invigorator plug-in. Pro Animator is a quick and easy 3D engine that can be run as a stand-alone application or an AE plug-in. The interface, while sporting a late-1990s-looking GUI, is a powerhouse with intuitive controls for bringing in 3D models from larger 3D packages, creating simple extruded models with controls that will be familiar to any experienced Adobe user.

Additionally Zaxwerks has 3D Flag, a quick way to make a 3D flag of any layer in AE. 3D Serpentine is another plug-in, designed to extrude 3D paths, which is great for making trails, film-strips, and ribbons. 3D Reflector will quickly generate reflections. 3D Layer Tools is an exciting tool for breaking up one layer into several without jumping to Photoshop and controlling them in Z-space.

Not all of these are essential, but they are all very useful. Pro Animator in particular is very valuable to most AE users interested in adding 3D. Unfortunately, even this far into the CS5's shelf life, there's no version of Pro Animator or 3D Invigorator that is compatible with CS5 for OSX. Luckily the stand-alone versions will work.

Trapcode Particular and 3D Stroke

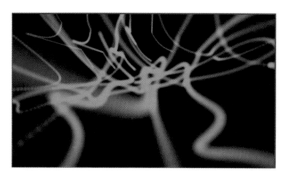 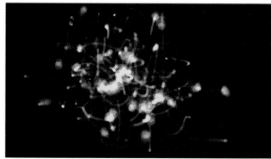

Figure 3.3(a) and (b) Trapcode plug-ins are all the rage in broadcast design.

A few years ago, the motion graphic images created on TV were literally dominated by Trapcode plug-ins (and images that looked like Trapcode plug-ins). The two most notable were the classic *Particular* and *3D Stroke*. Particular is a full 3D particle and dynamics package for After Effects. It is a wonderful tool, and if there are any plug-ins for After Effects that are a must-have, Particular is at the top of the list. Also, Trapcode makes 3D Stroke, which is kind of like the existing *Stroke* plug-in that can animate a line on a path, but 3D Stroke can do it on the Z axis.

Vector Image Editing

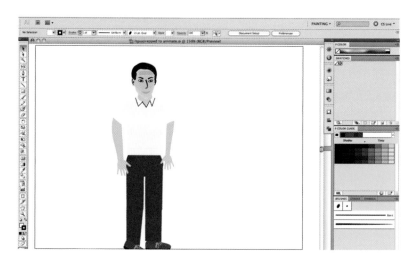

Figure 3.4 Illustrator.

Adobe Illustrator

It's optional for 3D but most motion graphics designers pretty much need to own Adobe Illustrator. After Effects works with Illustrator just like Photoshop: you use the software to supply AE with the artwork it needs to move. However, as a vector tool, you won't have to worry about the size of the image lessening the quality of the image. Another advantage of vector as it applies to 3D is that you can take the vector outlines from Illustrator and extrude them in 3D tools (most 3D hosts have an Illustrator file import function).

3D Software

3D Hosts

Although Photoshop and After Effects provide you with many tools for adding 3D to your motion graphics skill set, there's only a gamut of what they are capable of before you'd start to feel limited. A true 3D host will be needed eventually, and unlike with Photoshop and After Effects, the competition is fierce and numerous, and there's no clear sign that one software package is doing better than another.

Also, the core features are similar in that they provide tools for creation of 3D objects and animating them. Now, the degree and depth of the software varies, which we will discuss next. It's best for artists to either choose one and master it, or to pick up a variety of skills from different packages and play to the strengths of the software, which is cool, but will make you broke.

MAXON Cinema 4D

For this book we've chosen for the most part to stick to MAXON Cinema 4D for our lessons that require a 3D host. This was no arbitrary choice—it comes from several specific criteria. First, Cinema 4D's learning curve is considered by many to be

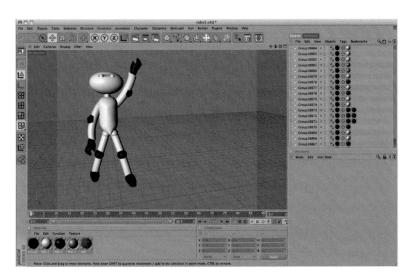

Figure 3.5

less severe than other 3D packages. Many motion graphics artists are really not looking to make Pixar-style full animation. Since they only need to add some 3D to their skill set, why not go with the one they most likely will pick up quickly?

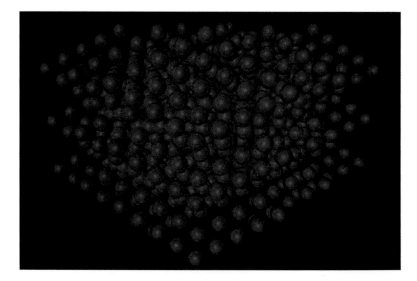

Figure 3.6 Compositing export from Cinema 4D.

Second, Cinema 4D has a famously smooth workflow with After Effects. You can save a multipass file of a Cinema 4D timeline as an .aec file that will render separate objects as layers for AE and preserve C4D's camera data, lights, and null objects. C4D supports various compositing software like Motion, Shake, Fusion, Combustion, and Final Cut. However, After Effects and Combustion have extra capabilities like the ability to edit cameras and lights, including animation.

The third reason for our choosing Cinema 4D is C4D's MoGraph tools. Anticipating the need for 3D in the motion

Figure 3.7 MoGraph tools are specific tools in Cinema 4D for motion graphics designers.

Figure 3.8 Cinema 4D's Cloner tool in action.

Table 3.1 Cinema 4D Pros and Cons

Pros	Cons
Integrates well with major compositing software	Folks who have used Autodesk 3D packages find C4D's interface awkward
MoGraph tools are a treasure chest for motion graphics designers	
Relatively quick learning curve	
BodyPaint is an excellent painting tool and it's included	

graphics market, MAXON introduced the MoGraph toolset. Designed to quickly and easily address typical motion graphics needs, they've developed an invaluable toolset.

MoGraph's first tool is the *Cloner*, which allows users to duplicate objects very quickly and set to predetermined patterns. There are MoGraph objects that are designed to create things like extruded text, displaced objects, and much more. *MoDynamics* allows users to adjust dynamic properties like mass, bounce, friction, and gravity. Also included with this toolset are *Effectors*, which are effects that can be added to MoGraph generators to quickly animate them.

Like its competitors, Cinema 4D is not perfect, but it is ideal for new 3D users to quickly add to their 2D skills.

3ds Max

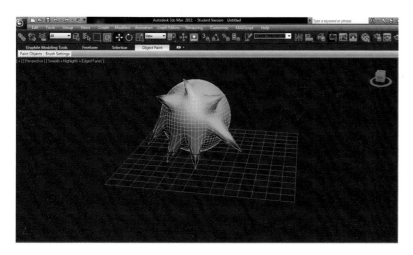

Figure 3.9 3ds Max.

3ds Max is a well-established tool of the trade. It's the most widely used commercial 3D software package. It's been around since 1990, and it is feature packed. It is not something that a book and a weekend can cure. Learning 3ds is a commitment of time and energy. The screen is literally littered with icons—pretty much everywhere there's real estate there is UI.

Among its most exciting tools are the numerous Modifiers that aid particularly in the modeling process. Many people who use 3ds came to it for its fantastic modeling abilities. Another tool that attracts a large amount of users is biped. Biped is a rigging tool that lets you use a prebuilt rig to shape into a character rather than forcing you to create a custom skeleton for each character you create.

Also there are great tools for architects and interior designers. One of 3ds Max's most winning features though is its speed on high-quality renders; this can really speed up workflow on many projects.

Table 3.2 3ds Max Pros and Cons

Pros	Cons
Fast renders	Long learning curve
Popularity means tons of web tutorials and resources	PC only; if you want to run it on a Mac, which is the common platform among motion graphics designers, you have to install boot camp
Large number of great modeling tools	If you want to integrate with After Effects you'll need a third-party plug-in
Biped makes for quick rigging	

Autodesk Maya

Maya is considered by many to be the most powerful 3D animation and VFX tool ever made and with good reason. Every film that has won an Oscar for best visual effects since 2001 has used Maya. The reason for Maya's strength in the industry is also the reason I've avoided covering it in this book: It is incredibly complicated, and a single individual is unlikely to master it in a lifetime.

Maya is known for its position as an entertainment industry standard. The reason for its rise comes from its ability to be customized. Maya can be completely reworked and custom fit to just about any need if you understand its built-in scripting language. Maya is also known for its modeling and UV mapping tools.

Table 3.3 Maya Pros and Cons

Pros	Cons
The entertainment industry standard	Long, long learning curve
An absolute powerhouse	
Flexible, can be used on any platform	Expensive

So, if you have the time and brain capacity, Maya's the way to go; however, if you'd like to dive in a little slower there's plenty of easier tools to learn out there.

Luxology Modo

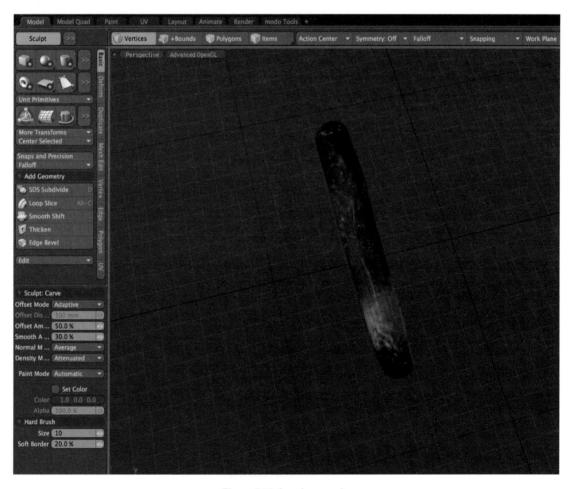

Figure 3.10 Luxology modo.

Relatively young in the field, modo has grown more and more impressive with each version. This package began as a dedicated modeler and grew into a full 3D package. It has a built-in paint and sculpture toolset with consistent tools. That's one of the advantages modo provides; learn a few tools and you are prepared for the rest of the software. Its animation tools are new, and there aren't any rigging features, but it's an absolute killer on the modeling/painting front.

Table 3.4 Modo Pros and Cons

Pros	Cons
Awesome modeling, sculpting, and painting tools	Not full featured yet: no rigging tools and it's obvious that animation features are new
Intuitive interface	
Full of useful presets	

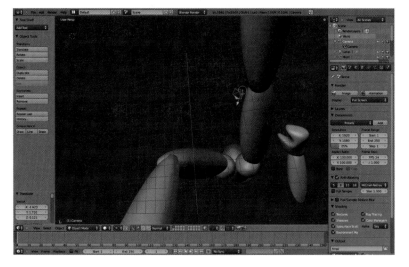

Figure 3.11 Blender.

Blender

Choosing a 3D software package is a commitment, based only on the cost factor it will set you back. Is there any way you can dive into the pool without spending a fortune? There is, though there is a little bit of a catch.

Blender is free. If you don't believe me, visit http://www.blender
.org/. Is it feature crippled? No, in fact it can do everything that
most major 3D packages can do. Does anyone use it? In fact,
because it's free it's probably the most widely *installed* animation
software. Although it's not exactly the industry standard, more and
more folks are using it on big projects in the entertainment and
game industries.

There are a few drawbacks. It has an odd interface; although
it can be picked up without too much suffering, it tends to scare
off some new users. Since it is free there's lots of documentation
to help new users. However being free and open source leads
to unfinished features, and some general buggy-ness. But with
open source those with the ability and the inclination can fix it
on their own.

Table 3.5 Blender Pros and Cons

Pros	Cons
Completely free, and you get the whole package	Interface scares off new users
Multiplatform, including Linux	Buggy

So even though I like the low-commitment and I encourage
people to download it, since this book is aimed at those who are
new to 3D, I would say that it's worth a trial but don't judge all 3D
software based on Blender alone.

Other 3D Host Options

These are just a sampling of the major 3D packages.
However, there are others worth noting. *Autodesk Softimage*
is another long-standing industry tool, which is PC only. It's
complex and has a long learning curve but if you are thinking
about the game industry its Face Robot tool is great for creating
quick face rigs. Another long-standing industry tool is *Newtek
Lightwave*. Popular on television, it does have a steep learning
curve.

Specified Tools

Among our specified tools let's start by looking at 3D solutions
that include content in order to speed up your process. These
tools are not required for many situations where a quick fix might
be needed, but it's worth looking into some of these tools.

Character Tools

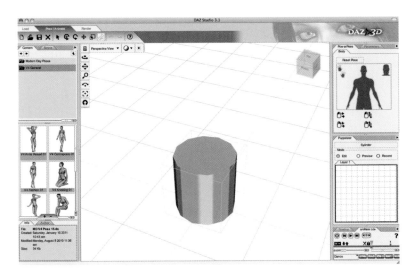

Figure 3.12 DAZ Studio.

DAZ Studio

DAZ Studio comes in two varieties: DAZ Studio itself is free and DAZ Studio Advanced is fairly low priced ($150) considering its competitors. DAZ Studio relies heavily upon providing users with premade content that can be manipulated to fit the needs of the user. Designed to approach a similar market to that of

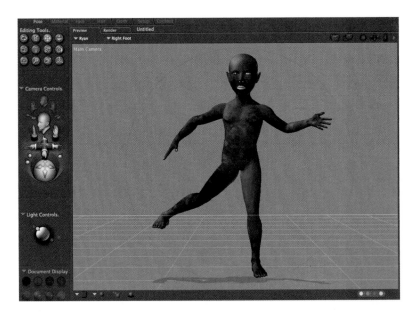

Figure 3.13 The quickest way to a 3D character is Poser.

Poser, DAZ Studio focuses more on the idea that users will want to purchase premade models rather than start from scratch. Also, the software comes with a basic feature set that can be augmented with the purchase of plug-ins that will expand its ability. DAZ will also take files from Poser.

Smith-Micro Poser

Poser works in a similar vein to DAZ Studio, supplying users with content that can be manipulated for use in their own projects. So Poser comes with libraries of included content, but more can be added. Poser is also well liked for its ease of use, even for beginners. Poser's greatest strength is in the area of creating 3D figures that will be part of either an animation or illustration. It can make quick work of modeling, texturing, and rigging.

Environment Tools

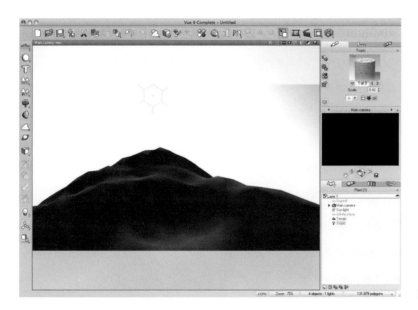

Figure 3.14 Terraforming in Vue.

e-on Vue

Vue is designed to make quick work of 3D landscapes and scenery. It's a full scene generator, with controls over things like vegetation, wind, and light. It's meant to be like an environmental artist in a box. It comes in a variety of packages designed to be very user-friendly, and works nicely with Poser. Also it is available as Ozone, a plug-in version of Vue, which works with Cinema 4D, Maya, 3ds Max, Lightwave, and Softimage. Ozone is not the full version of Vue; it's restricted to atmospheres and skies.

Digital Sculpture

Digital sculpture is a newer modeling technique that has only been around the industry for a short time but has already gained a very large number of devotees. Rather than traditional 3D modeling techniques, digital sculpture tools allow users to use their mouse or Wacom tablet to push and pull on meshes until they resemble very real surfaces. These tools have paved the way for high poly models that can look photorealistic.

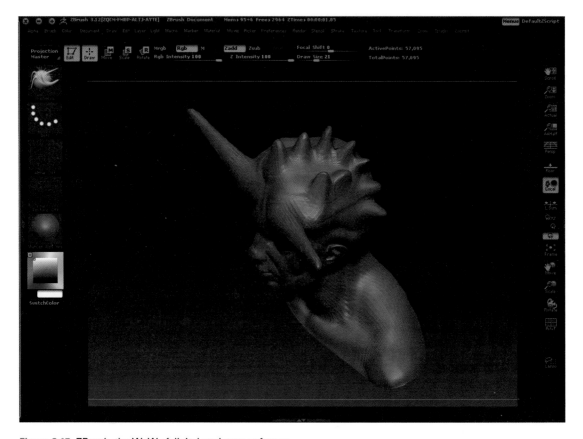

Figure 3.15 ZBrush, the WoW of digital sculpture software.

Pixologic ZBrush

ZBrush has gone from a homemade custom tool to nearly an industry standard in a short period of time. It enables users to manipulate 3D models with intuitive sculpture-like controls, rather than the standard vertex manipulation tools more commonly found in the major 3D hosts. At the heart of this software is a technology referred to as *pixols*. Basically, a pixel contains color and x and y position data. A pixol adds data for the depth or the position of the pixel in Z space. Additionally exciting for ZBrush users are the painting features, which allow users to paint on their models without texture mapping beforehand.

ZBrush does have its detractors. Its interface, though highly customizable, is like nothing else out there and it is rarely thought of as being intuitive. It's also designed to be a self-contained solution, which Pixologic is getting away from (by offering the GoZ feature in version 3.2, it now can talk to other applications). However its self-contained nature has led some of its users to the competition.

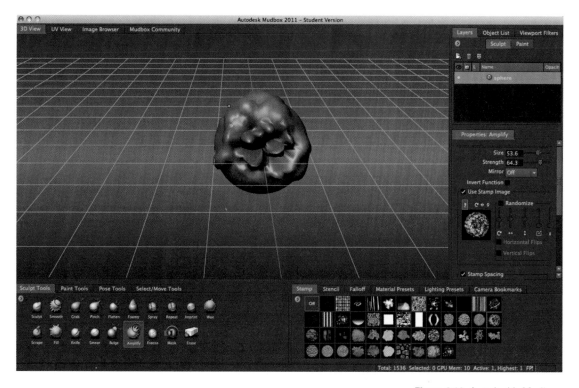

Figure 3.16 Autodesk's Mudbox.

Autodesk Mudbox

Mudbox was introduced essentially to be an Autodesk version of ZBrush. Seeing that the industry was moving in this direction, Autodesk figured that it was a good opportunity to create a tool that would integrate smoothly into their current lineup of software. Mudbox is also pretty easy to use, especially when compared to other Autodesk software—who would have expected them to release the more intuitive software than another publisher?

Users tend to be divided when trying to choose between the two packages. This is the stuff of major flame wars in the forums. Some say ZBrush's odd interface put them off right away, while

Mudbox's comparable simple interface and easy workflow make it an obvious choice. Others prefer the robust features that ZBrush's deep toolset offers.

Dedicated Modeling

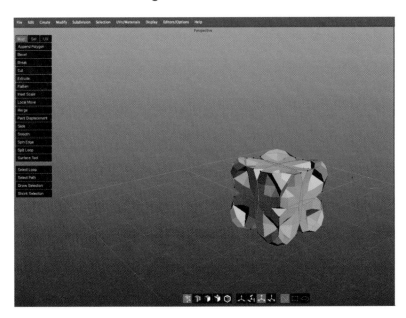

Figure 3.17 All the tools you need to model in a quick, intuitive package.

Nevercenter Silo

As mentioned earlier, there is a lot of bloat in 3D software. Also most major 3D packages are a good decade or two old. So Nevercenter introduced a dedicated modeler and in doing so created a great entry into 3D modeling. Silo has every necessary tool for modeling in an accessible location. It's amazing how fast anyone with just a little modeling experience can be up and running. The new workflow of base geometry-sculpture/texture-animation in separate software has created a perfect niche for geometry tools like this. Plus, it is very reasonably priced.

Mobile Tools

Whether a serious 3D tool can be found on a mobile platform is debatable. But I often feel like modeling when sitting in the doctor's office waiting room. Apple's mobile operating system, iOS, has everything under the sun in the App Store, and there's a handful of somewhat useable 3D modeling tools that work well for showing a model on a mobile platform. Some do well with just a few modeling features.

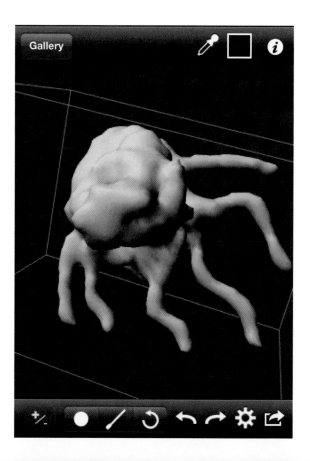

Figure 3.18 (a) and (b) iOS digital sculpture tools, Sculptmaster 3D and iDough.

Sculptmaster 3D and iDough are two examples of digital sculpture–style tools on the iOS platform. While expecting an iPhone ZBrush will lead to some disappointment, the whole idea of digital sculpture with a touch interface makes a great deal of sense to me. Sculptmaster3D is extremely intuitive, and although it's handy, it leads all too often to the 3D equivalent of finger paint, which may be a cool thing in its own right, but has limited application. iDough, though a little more powerful, is a bit more difficult to get used to.

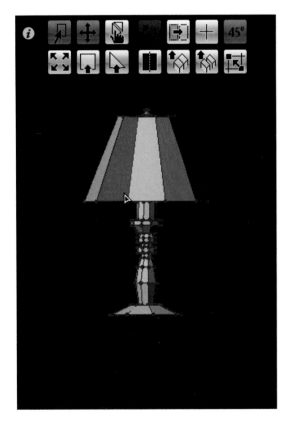 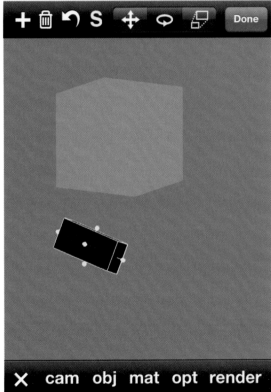

Figure 3.19 (a) and (b) iSculptor and iTracer are more traditional vertex modeling.

iSculptor and iTracer are two examples of more traditional 3D modeling tools. iTracer in particular feels very well made with the iOS format in mind. iSculptor is not quite as nice and it's the only time I've seen a mouse-styled pointer on the iOS platform. However it comes with lots of premade, handy models to use as a start point.

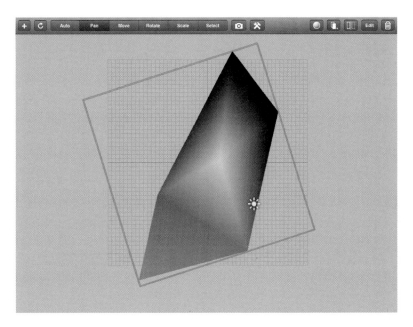

Verto Studio is the most full-featured 3D modeling app I've seen on iOS, which might explain why it's iPad only. It's in the vein of traditional 3D modeling tools, and it's cleanly done for the iPad.

PHOTOSHOP 3D

The process of creating 3D objects is called *modeling*. In 2D digital art, we have the option either to take a photograph or to draw the object needed. With a 3D object we have to do a few more things. Before we can concern ourselves with the texture of an object, we have to define its shape. So in order to define a shape, we start by working with what are called *primitives*.

Even if you have never worked with 3D software, you are probably familiar with primitives. They are basic geometric objects that have three dimensions. A cube, sphere, pyramid, and cylinder are all examples of 3D primitives.

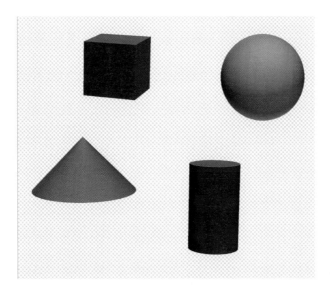

Figure 4.1 Four 3D primitives from Photoshop.

As we get further into 3D modeling we will be able to create most of our objects from modifying and combining primitives.

At the end of 2008, Adobe released Photoshop CS4, which boasted the capability of being able to apply any Photoshop layer to a 3D primitive with a few quick clicks. In 2010 Adobe introduced CS5 with more features, including, most notably for 3D artists, the Repousse tool. In addition to that, within Photoshop, you even have the ability to add keyframe animation in 3D space (or take it over to After Effects CS5 and do some more complex animation).

Are Photoshop and After Effects, the long-term solution for most 2D animation concerns, now a full 3D solution? Well, no, or at least not really. For designers who are Adobe based, it offers some nice extensions to what your software is already capable of handling. However, complex 3D requires tools with much deeper capability.

Throughout this book we will be often addressing the issue of how much 3D you need. Many motion graphics designers will not need to create fully articulated 3D characters; a common need is something like a 3D product shot. The Photoshop/After Effects CS5 solution will work quite well for that.

You can't model complex surfaces; mainly there's no feature for working with NURBS. NURBS is a function of 3D modeling and animation software that gives you Bezier controls of the surface of an object.

In addition to not having NURBS, Photoshop also limits the complexity you can create by forcing you back to a 2D layer when you try to combine multiple 3D objects. However, even with these limitations, the new 3D feature set in Photoshop CS4 is pretty powerful and can be quickly learned by most Photoshop users.

In the following tutorial we will create a simple 3D object in Photoshop. We're going to start from a photo of a button, and turn it into a full 3D object.

Creating a 3D Button in Photoshop

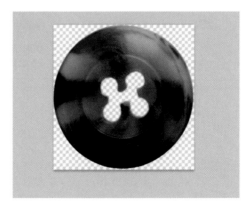

1. Open the *button.psd* image in Photoshop. Our document is set to a square size of 500 px × 500 px. This makes doing what we are about to do a little easier.

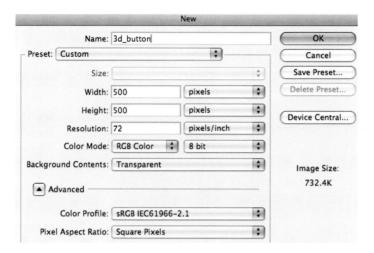

2. Create a *new document.* Make it the same size as our button.psd.

3. Fill our new document with gray. Rename the layer "button."

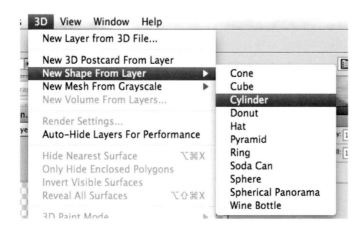

4. Go to the *3D* menu and choose *New Shape From Layer>Cylinder.*

3D Tools in Photoshop

3D Object Tools (K)

3D Camera Tools (N)

Figure 4.2

Before we go any further with this tutorial we should familiarize ourselves with the 3D tools. With the arrival of version CS4, Photoshop introduced two brand-new tools for dealing with the new 3D functionality. (I know for most software, two new tools are nothing exciting, but a change to the iconic Photoshop toolbar is a pretty big deal.)

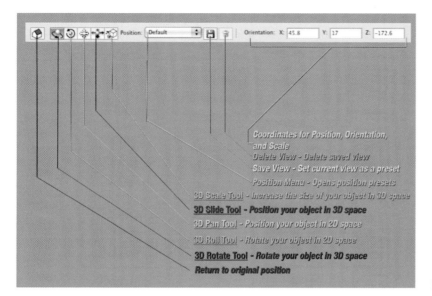

Figure 4.3

First we have the *3D Object Tools*, which are meant to move, turn, and manipulate 3D objects themselves. Figure 4.4 demonstrates what each one does. To get to the *3D Object Tools* from the keyboard press K; to cycle through them press Shift-K. They are also displayed on the top options bar.

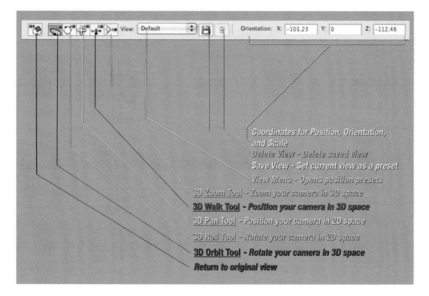

Figure 4.4

The other new tool is the *3D Camera Tool*. The functionality of this tool will feel quite similar to the *3D Object Tool*. The major difference will be that you are not changing the object itself, but you are adjusting and manipulating the view of the object from a virtual camera. Figure 4.5 shows all the *3D Camera Tools*, which can be reached by pressing the N key on the keyboard, or you can cycle through the *3D Camera Tools* by pressing Shift-N.

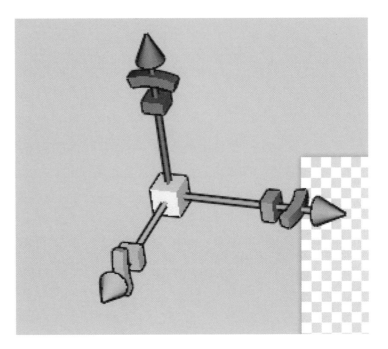

Figure 4.5

In addition, we also have the new *3D Axis* tool; whenever you have an object that is a 3D layer and you are using a 3D tool, it will appear. Many of the 3D editing tools can be accessed from here. Each arrow represents a specific 3D axis. The red arrow is the *X axis*, the green arrow is the *Y axis*, and the blue arrow is the *Z axis*.

If you place the cursor over the tip of the arrow it will turn into the *3D Slide* tool; however you are only moving the object on the axis that corresponds to the arrow you are on.

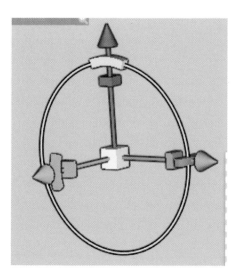

Directly below the arrow you will see a polygon shape that when highlighted will reveal a ring that represents the orbit of the object you are on. Since we are working on just the corresponding dimension of the arrow we are on, it will active the *3D Roll* tool, and you are able rotate on the selected axis.

Right behind the rotate tool discussed earlier you'll see a cube on which you can use the *3D Scale* tool to stretch or squash the shape on the selected axis.

The cube in the center allows you to use the *3D Scale* tool to do a proportional scale of the entire 3D object.

TIP

Although the 3D Axis tool is very useful, in the midst of a design it can be a little distracting. To tuck the 3D Axis tool away or resize it, use the arrow in the upper left-hand corner of the document.

The magnifying glass can be scrubbed to scale it.

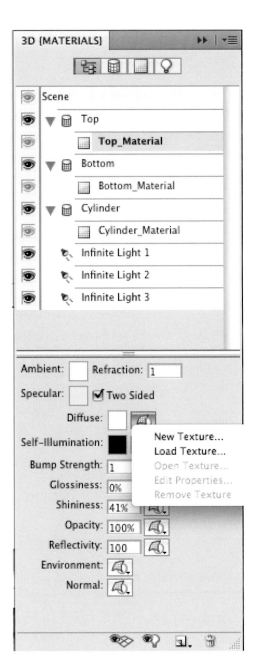

To resume our tutorial:

5. Go to the *3D Materials* panel. Select the *Top Material*. Go to the *Diffuse* setting toward the bottom of the menu and click *Load Texture*. Choose the *button.psd* file. Repeat the same process for the *Bottom Material*. Also, change the color setting ▪ next to where we chose our texture for *Diffuse* to black.

The 3D Panel

The 3D panel in Photoshop is how we will interact with most of our 3D objects. Here we can do a number of things, such as assigning and editing textures. It's divided into four subpanels. Figure 4.6 provides an explanation of each subpanel.

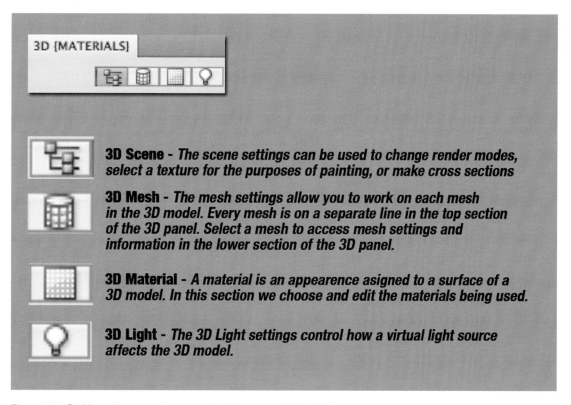

Figure 4.6 3D objects have complex controls that can be adjusted in the subsettings of the 3D panel.

6. Now that we have assigned the materials we have the problem of our button being a long tube shape. In order to bring it back down to its correct size, we'll use the *3D Scale* tool. Go to the top options bar and reduce the scale on the Z axis.

7. Now you can switch back to the *3D Rotate Tool* to give your new button a quick spin!

Animating 3D Objects in Photoshop

Though Photoshop's main purpose for existence is not exactly for animation, the Extended Version of CS5 has a full keyframe animation timeline just like After Effects (with far fewer options). When Adobe added all the wonderful new 3D options, they included them in the Animation window. So in the following tutorial we'll make our button spin.

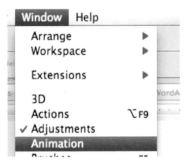

1. Open the *Animation* panel by going to *Window* and choosing *Animation.*

2. Make sure that you are viewing the *Animation* panel in *Timeline* view and check to make sure that the word Timeline appears in parenthesis next to the word Animation. If not, you are looking at the timeline in *Frame* view; click the switch button in the lower right corner of the panel.

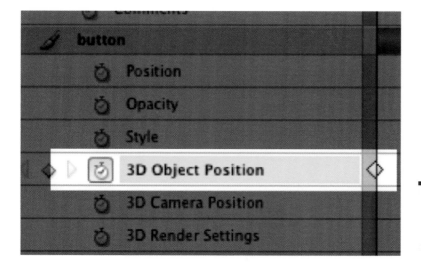

3. The *Animation* (*Timeline*) panel will recreate your layers panel, and when you want to animate a layer you will need to target the layer and click the twirly arrow to open the animation tools. *Position, Opacity,* and *Style* are the standard 2D space animation tools. On 3D layers you have four more animation tools; for this lesson we are going to focus on *3D Object Position*.

4. To begin an animation, just like you would with Adobe After Effects you click the *stopwatch* icon for the animation tool you'd like to use. Since it is our plan to animate the *button* object's rotation, click the stopwatch for *3D Object Position*. Just like in After Effects, if you click the stopwatch again it will erase all your keyframes for the animation tool being used.

TIP

Always remember, when you are using a keyframe animation software, you need at least two keyframes to have an animation. The software needs to know a start point and an end point so that it knows the values for the frames that it has to tween.

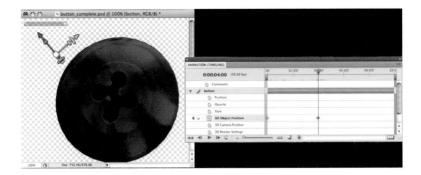

5. Move the *Play head* to the 4:00 mark on the timeline. Go to the *3D Rotate Tool* and turn the button. The software will automatically give you the second keyframe. Now you can press the spacebar or *Play* button and you will see the button move.

Creating a More Complex 3D Model in Photoshop

Although Photoshop can't create as complex a 3D model as a dedicated 3D application, it can do more than slap textures onto primitives. We can also use the *3D Mesh From Grayscale* function. This will allow us to use the common 3D principle of the *bump map*. A bump map will extrude a flat object in 3D space based on grayscale levels.

Figure 4.7

Figure 4.8

In the image you'll see first a gradient that goes from black to white. Then *3D>3D Mesh From Grayscale>Plane* was applied. Now you'll see, moving from left to right, that the plane rises according to grayscale level. Below you'll see a more complex gradient and what happens.

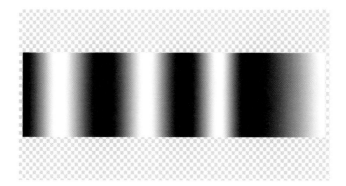

Figure 4.9

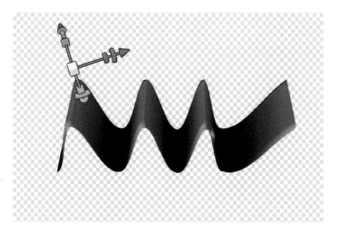

Figure 4.10

As I am sure you are starting to see, there's a lot of excellent potential for using *3D Mesh From Grayscale*.

Creating a 3D Flower Using 3D Mesh from Grayscale

1. Make a new Photoshop document. Use the *NTSC DV* preset under *Film and Video*.

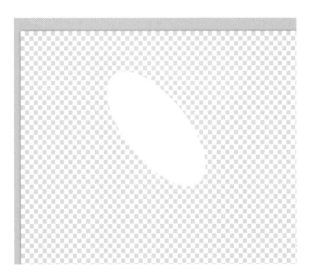

2. We are going to create an oval with the *Ellipse Tool* (it's under the *Rectangle Tool*).Use the *Fill Pixels* mode 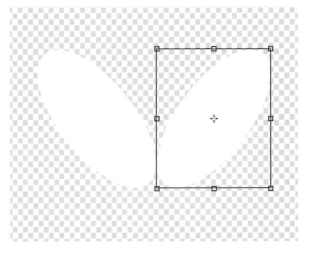 in the options bar. Create your oval on a diagonal by clicking and dragging from the upper left to the lower right.

3. Duplicate (Cmd-J) the layer with the oval. Press V to switch to the *Move Tool*. If you don't have *Show Transform Controls* activated, do so by clicking the checkbox next to it in the top options bar. Grab the *Transform Controls* from the left-hand side and drag it over so you have something like what I have in the above figure. Apply the transformation. Highlight the two layers and press Cmd-E to merge the two layers.

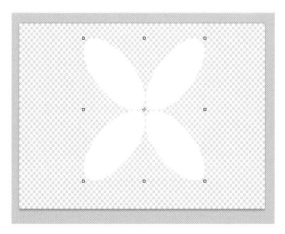

4. Duplicate the layer and use the *Transform Controls,* this time clicking from the top of the box and dragging straight down. Apply the transformation. You should now have something that looks like in the above figure.

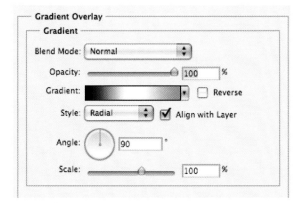

5. Apply the *Layer Style* for *Gradient Overlay* to the image. In the Gradient Overlay dialog box, switch the style to *Radial.* Click the gradient and change it to look like the one in the above figure by

TIP

 To get a more odd-looking orchid-like flower, try experimenting with the gradient pattern you make here. One thing to try is to have the black marker in the middle and white on the end; you'll get a more unusual shape.

moving the white marker to the middle, and add a gray marker at the end. To add a marker click below the gradient bar and choose a color. Click *OK*.

6. Apply *3D>New Mesh From Grayscale* and choose *Sphere*.

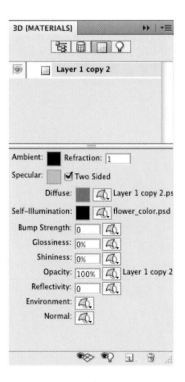

7. There are a few more settings to tweak here. We want the surface of our flower to have the right feel and texture. Nothing calls out bad 3D modeling like inappropriate textures. So go to the *3D Materials* panel, choose *Self-Illumination*, and edit the texture setting by loading the flower_color.psd document from the DVD.

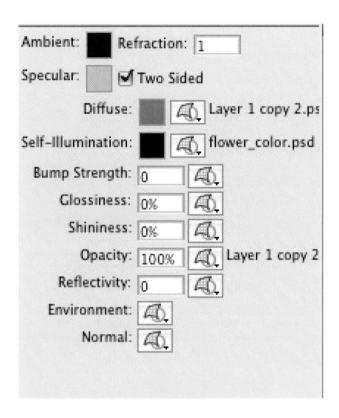

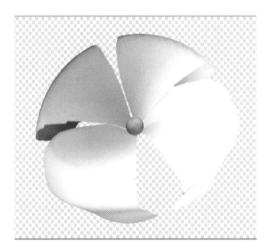

8. Next turn your attention to the settings below *Self-Illumination.* The kind of flower we are going for here doesn't have a glossy surface so lower the *Glossiness* setting to 0%. Also, lower *Shininess* to 0%. To make the little bud in the middle, I added a new layer that I painted yellow, and I used the *New 3D Shape From Layer* option from the *3D* menu. Then I used the *3D Scale* tool to make it small enough to fit nicely in the center.

Creating a 3D Scene in Photoshop, and Using After Effects for Animation

For as long as I've been using After Effects (I started with version 4 back in 1999), folks have been pining for more 3D capabilities. After Effects 9.0 in the CS4 suite is the largest step it's taken in this direction since the product was first introduced. Later in this book we look at the amazing new functionality where it allows you to import a 3D model from Cinema 4D or other 3D software packages. For right now, we'll get to know the basics of 3D in After Effects by going through the process of creating a full 3D scene from primitives in Photoshop and importing and animating them in After Effects.

Creating Textures

To begin building our 3D scene, we need to make some 3D objects and backgrounds. To start making our planets, we'll begin with painting a planet surface. There are many ways to make textures in Photoshop. In this section we'll use one widely known technique to create a surface that will look like terrain.

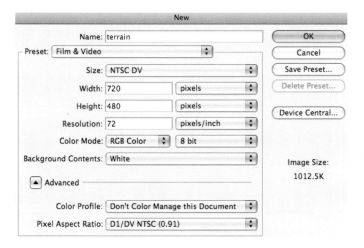

1. Make a new Photoshop document and choose the *NTSC DV* preset from the *Film and Video* list under presets.

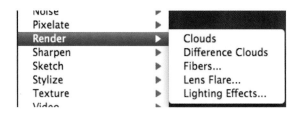

2. Choose two terrain style colors; I'm using a light blue and dark green, but it doesn't really make a big difference which colors are used as long as one is fairly bright and the other is dark. Choose *Filters>Render>Clouds*. You should have something like the above image.

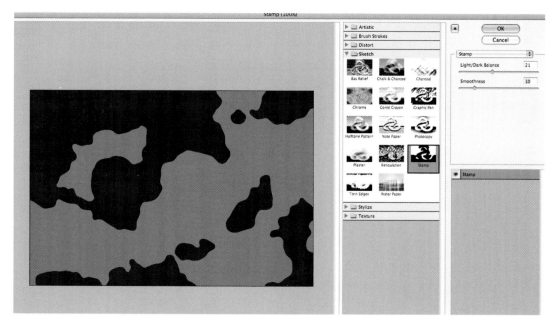

3. Open the *Filter Gallery*. Open the folder called *Sketch*. Apply *Stamp*. Adjust the *Light/Dark Balance* until you have landscape-like shapes. Then click *OK*.

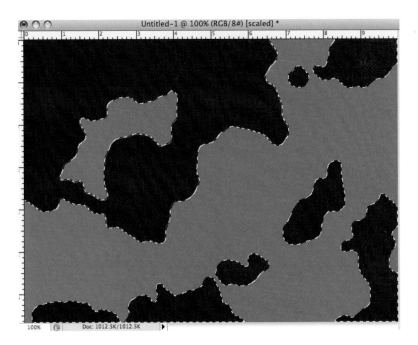

4. Use the *Magic Wand* and select one of the "continents." To select all of our terrain, go to *Select>Similar*. Now you should have all of your fake land selected.

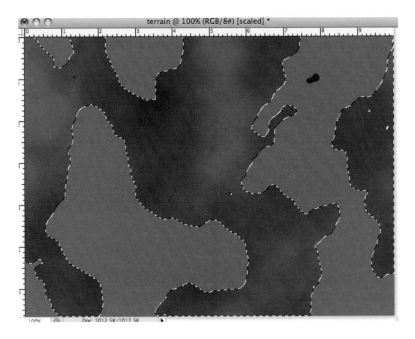

5. Choose *Filters>Render>Clouds* while you have your terrain selected. This will create the cloud effect inside your selection.

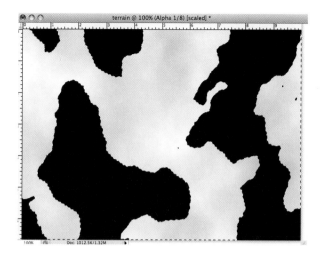

6. Go to the *Channels* tab. Click the *New Channel* icon to create a new *Alpha Channel*. Now apply *Filters>Render>Clouds* again, which will create something that looks like the above figure. Now click the *RGB* channel.

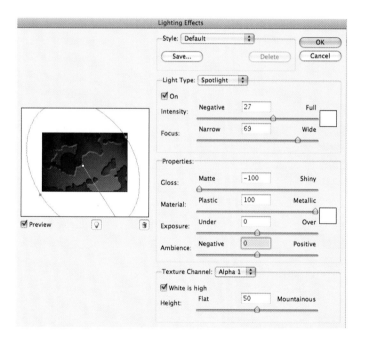

7. Press Cmd-D to *Deselect*. Now go to *Filters>Render>Lighting Effects*. Go to the *Texture Channel* and choose our *Alpha 1*. This is the alpha channel we just made. Set the *Light Type* to *Spotlight*. Move down to the *Properties* section. Put the slider for *Gloss* all the way down to *Matte*. Feel free to try out different settings for the other effects until you are happy with the look and then press *OK*.

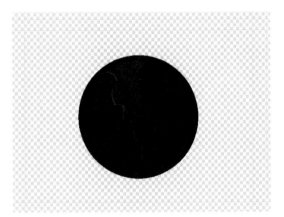

8. Now that we've got some terrain we can apply *3D>New Shape From Layer*. Pick *Sphere*, and now you have a planet that we can use as part of our space scene. Repeat this a couple times for each planet you need.

Space Background

This is a quick Photoshop lesson that's a bit of an aside from 3D techniques, but it's a handy thing to know and it's going to be a big part of our space scene. Here's how you can generate a space background in Photoshop.

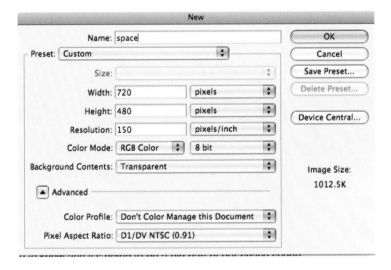

1. Make a new Photoshop document. Since this is going to be a background for a 3D scene, use a high resolution. I've chosen 150 ppi. It will also help the filter we are going to use generate more pixels to work with at a higher resolution.

2. Go to *Edit>Fill* and choose *black*. Rename your layer black. Make a new layer, go to *Edit>Fill* again, and choose *50% Gray*. Rename this layer stars 1.

3. On the stars 1 layer, we are now going to apply *Filter>Noise>Add Noise*. Set *Amount* to *100 percent*. Select a *Uniform* distribution and check *Monochromatic*.

4. Apply *Filter>Blur and Sharpen>Gaussian Blur*. Keep it very low, around 1 pixel.

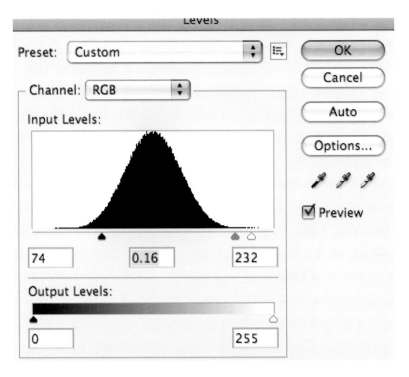

5. Now we can turn this mess of blurred static into something that looks like stars. Go to *Image>Adjustments>Levels* and move the sliders to about the positions you see in above figure. The *mid-tones* slider has the most control; adjust until you've got the amount of stars you need.

6. Use the *Blur* tool and blur some of the star clusters to give the star fields some depth.

7. From here forward is totally optional. If you are happy with what we've done so far, feel free to stop here. However, if you feel the need to make our space scene a little prettier, oddly colored clouds will add to that otherworldly feel. Create a new blank layer, rename it clouds, and use *Filters>Render>Clouds*. Then apply *Filters>Blur* and *Sharpen>Radial Blur*. Choose a *Zoom* blur type. Set the Blur's center to wherever you'd like, and make the *Amount* fairly high, around *80*. You should have something along the lines of the image in the above figure.

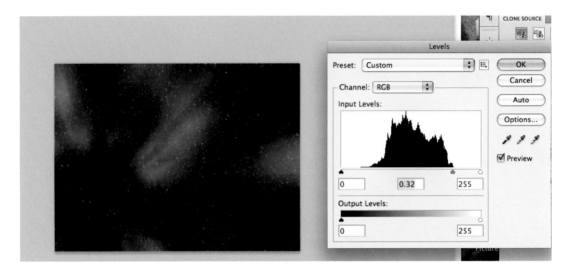

8. Set the clouds layer's blend mode to *Screen*. Then use *Image>Adjustments>Levels* and move the *mid-tones* slider to the very edge of the highlights side of the histogram.

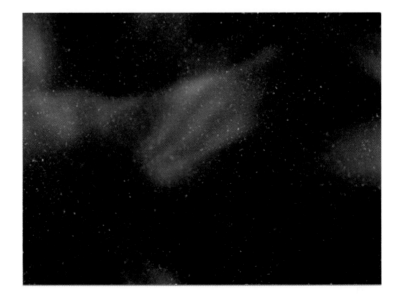

9. Use *Image>Adjustments>Color Balance* to give the new clouds some color. If you'd like more than one color for your

space clouds, duplicate the clouds layer, go to *Image>Adjustments>Desaturate*, and then apply *Image>Adjustments>Color Balance.* You may need to reshape it so it's not exactly the same shape as the existing cloud; try using the *Edit>Transform* tools to manipulate it. If you need to eliminate certain areas, you can paint them with black. Remember that the *Screen* blend mode eliminates the darkest colors, and painting with black can be like using an eraser.

Animating a 3D Scene in After Effects

After Effects CS4 can take 3D layers from Photoshop and animate them in real 3D space. The controls will differ somewhat from the normal run of things in After Effects so here's how it works.

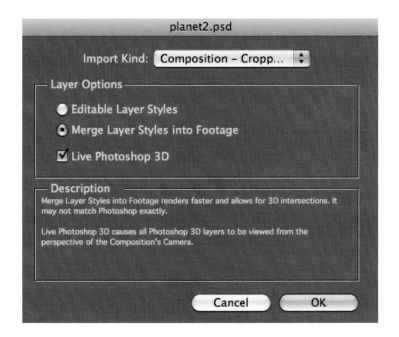

Figure 4.11

TIP

In addition to .psd files, you can also import the following 3D file formats into After Effects:

.3ds (3ds Max)

.dae (Digital Asset Exchange, COLLADA)

.kmz (compressed Keyhole Markup Language format, Google Earth)

.obj (common 3D object format)

.u3d (Universal 3D)

Since we are focusing on Cinema 4D for our modeling in this book, Cinema 4D can export the 3ds format. Also, the freeware Google SketchUp modeling software can export in the .kmz format.

When you import a 3D Photoshop file, be sure to import it as a *Composition* or *Composition Cropped Layers* because importing it as *Footage* will flatten the document and discard the 3D settings. After you've given the import command it will bring up the .psd import options dialog box. If you've used *Layer Styles* you can choose whether to leave those editable or merged to their respective layers, and you also have the option of using the *Live Photoshop 3D* settings, which directs all Photoshop 3D layers to be viewed from the perspective of the After Effects composition camera.

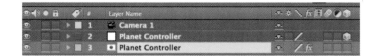

Figure 4.12

A 3D object in After Effects will actually be three layers. First, each 3D object has its own *Camera* layer. The second layer is a *Null* layer that is used to control the layer content. Finally the actual imagery itself is in the bottom layer. You will control your layers for the most part from the *Null* controller layer. In the next tutorial I'll take you through animating a 3D scene in After Effect CS4.

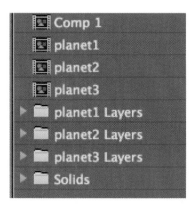

1. Setting up a 3D scene with multiple 3D objects is kind of a task with AE CS4, something I hope will be fixed in future versions. For right now though, let's import our planet documents, one at a time. Use the setting for *Composition Cropped Layers* and make sure that *Live Photoshop 3D* is activated.

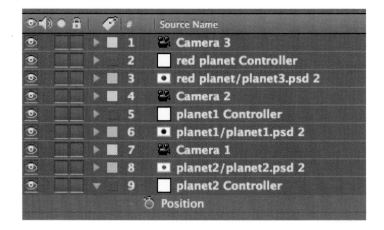

2. Make a new *composition*, name it spacescene, and use the *NTSC DV* preset. Go to the three separate planet compositions

and copy and paste the three layers contained in each composition into our spacescene. Be careful, and match what I have in the above figure.

3. Now let's import our space background. When you have it in the project drag it into our timeline at the bottom of the layer order.

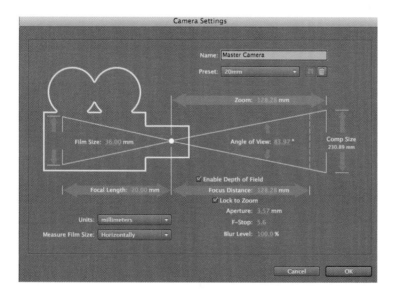

4. We'll now create a camera to animate a fly-through of our scene. Create the new camera layer by going to *Layer>New>Camera*.

Name the new camera Master Camera. Use the *20mm* preset and then click *OK*.

5. To animate our planets we'll go to the planet's controller layer. Highlight the *planet1 Controller* layer and press the R key to open its *Rotation* tool. Press the *stopwatch* icon to begin rotating the planet. We won't change any number values here, but toward the end of the timeline change the *Z Rotation* value to 1 × 0.0 to give it one revolution. Repeat this on the other planet layers to get each of them rotating.

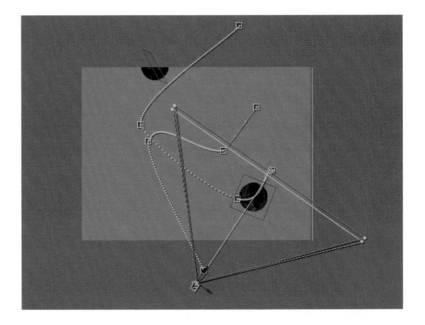

6. Recalling what was discussed in *Chapter 3* about *Camera Layers* in After Effects, animate the camera through our space scene by keyframing our Master Camera layer's *Position* and *Point*

of Interest. In the above figure, I switched to the Top camera view and keyframed its *Position* weaving through the planets. I adjusted the *Point of Interest* to turn it toward each planet as it flies through.

Repousse

Repousse is an extrusion tool that debuted in Photoshop CS5 and is a long-awaited feature, especially for those of us who have been wanting a quick way to make animate-able 3D type.

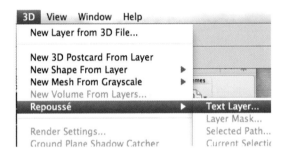

Figure 4.13

To get started with Repousse we use a type layer; choose *3D>Repousse.* Choose *Text Layer....*

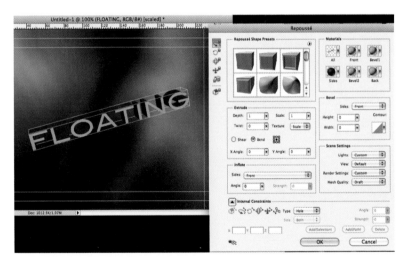

Figure 4.14

When using Repousse you can begin by choosing a shape for the extrusion from the *Repousse Shape Presets*. You can adjust the depth and scale, and even add twist to the extrusion. You can set different materials for the different angles under the *Materials* section. Under *Bevel* you can change the shape of the edge contour. If you want to bloat the front surface of your type, the *Inflate* section has options for that.

Interview: Luc Dimick

Figure 4.15 Lucas Dimick, painter and animator.

Lucas Dimick is an artist and an award-winning independent animator. He works in a wide range of media from paint on canvas to hand-drawn and computer animation.

As well as making art, Lucas has also worked professionally as a graphic and motion graphic artist in New York and Denver. Lucas's animations have been screened at The Big Muddy, Boston Underground, and Chicago Underground film festivals. His 23-minute animated film, *Prison Beta*, won Best Animation at the 2008 Chicago Underground Film Festival.

Lucas received his Masters of Fine Arts degree in Film, Video, and New Media from the School of the Art Institute of Chicago (www.saic.edu).

He currently lives and teaches in Austin, Texas.

What compelled you to start a career in art?

It was the only thing I was good at as a kid. I remember my mom saying, "Luc, you are pretty good at this art thing, you should be a graphic designer, that's the only way you'll make any money."

What was the first piece of animation software that you used and why?

The first animation software I used was Director, on the job. Director was kind of one of those programs we had on the hard drives that we never used, but I would play around with it.

What do you use now?

I'm using primarily After Effects and Flash.

Can you tell me what your workflow is on a project?

I start with frame-by-frame animation because I like the squiggly-line effect of the hand-drawn look. From Flash I'll export that out and bring it into After Effects and composite it with backgrounds I've created in Photoshop, Illustrator, or other .swf files. I'll typically layer it in that way.

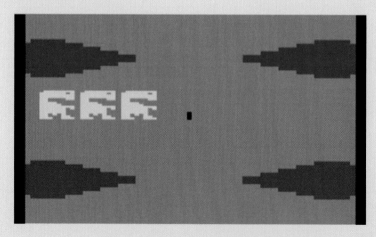

Figure 4.16

Do you make use of the 2.5D aspects of Flash and After Effects?

Yes; for example on my film *Prison Beta* I used it to bring another layer of dimensionality to my 8-bit characters to make them look like they are turning in 3D space. Rather than just kind of flipping in 2D space, now I use it when compositing swfs in After Effects to get that extra sense of depth and perspective in between the foreground and background.

Can you share a moment when you found something really inspiring?

I always kind of fostered this interest in the fine arts and kind of kept it hidden away and focused on graphic design and animation because this kind of art is 'okay.' This other kind of thing, fine art, is for people who live under bridges. Until I saw *Basquiat*. That film inspired me and made me really excited about fine art, it made me think this is something I can do and I don't have to be ashamed of it. Once that door opened I saw how I could combine that interest in fine art with my skills in graphic design and animation.

What aspect of the 3D skillset appealed to you as an artist with a background in traditional media and 2D animation?

I really like that technique I've seen in *The Triplets of Belleville*, where it's obvious that a 3D-generated element was added, like the vehicles, but had a 2D feel to it. I'm interested in that area where 3D can overlap into 2D.

Where do you look for inspiration?

Other artists, the work of other artists. Recently it's been Cy Twombly. My interest in him comes and goes, and recently it came back and I was looking at his work again. He passed away recently. I'm interested in that scratch and smear style of painting and making it move, animated. It would be interesting to see how 2.5 could play into that, and get that depth I was talking about earlier.

What is your dream project?

An animated feature of some sort. Something nontraditional. Nontraditional meaning something more in the vein of *Triplets of Belleville* than say something like *Bakshi*.

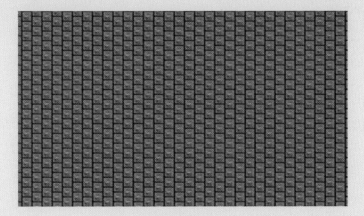

Figure 4.17

What have you done so far that you are most proud of?

My short *Prison Beta*. It's a 23-minute found-footage/8-bit animation short film. It's not the animation that I'm most proud of, it's the conceptual idea behind it.

Continued

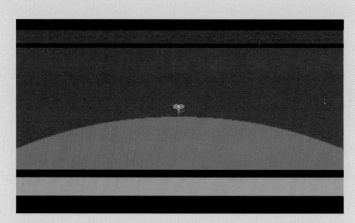

Figure 4.18

What skills do feel like you are still learning or improving upon?

I've been spending a lot of time working on my skills as a traditional 2D animator and translating my drawings into animation. it's good that has been able to cross over nicely into my teaching. The students in my 2D classes tend to gravitate towards a more hand-drawn look.

I feel like college-aged students right now have a real appreciation for craft, do you think that 3D has had an influence on this?

Oh, definitely. 3D has set this bar for quality. That bar has been set, and they are working to get their 2D stuff to that level of quality.

Is there any movie that you've seen that you wished you worked on?

Ratatouille, my favorite Pixar film. *The Triplets of Belleville* and *The Illusionist.* There's a lot actually.

DESIGNING FOR DEPTH: ILLUSTRATOR 3D TECHNIQUES

There are a number of techniques that can be used in Adobe Illustrator to add depth to your work. Is Illustrator an application like Photoshop with tons of tools designed to go to 3D software? No, not at all, but there are plenty of uses that we will discuss throughout this book. First, Illustrator allows us to quickly outline text, which allows us a great deal of creative control of our type. Second, the vector line art can jump from Illustrator to a 3D host quickly and painlessly.

The Advantage of Illustrator in After Effects

The basic advantage of using Adobe Illustrator in After Effects over Photoshop comes down to a couple tools. Illustrator is a *vector* image editor. Vector images don't assign pixels to specific locations; rather they save *Bezier* shape data, so the shapes can be resized without losing image quality. The downside is that vector images can't create a photographic image (in most cases; some artists use techniques to make Illustrator look nearly photographic but in general this is not what Illustrator does best).

TIP

Vector and raster techniques are employed when working with 3D models; vector tools are used to determine the size of a shape, and raster images comprise textures. Understanding the difference between these methods of image making will lead to greater success with 3D software.

Figure 5.1 Illustrator layers can use the Continuous Rasterize switch in After Effects so there is no distortion on the Scale tool.

When you import an Illustrator layer into After Effects, you can use the *Continuous Rasterize* switch in After Effects. This tells After Effects to go back to the Illustrator file and rerasterize at the larger height. This will keep a very crisp image quality.

3D Tools in Adobe Illustrator

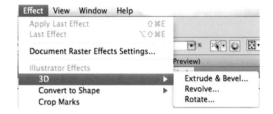

Figure 5.2 3D tools in Illustrator.

The 3D tools in Illustrator have been there a long time, and can be quite useful. In *Effect>3D* you'll see *Extrude & Bevel,* which is a standard extrusion tool that will extend edges in 3D space.

TIP

Illustrator's other 3D tools, *Revolve* and *Rotate,* can be applied for similar techniques as the one used in the following tutorial, but *Extrude* and *Bevel* work best for type. *Revolve* will work nicely for objects that would need a more ceramic style shape. *Rotate* is a 3D rotation for a flat object.

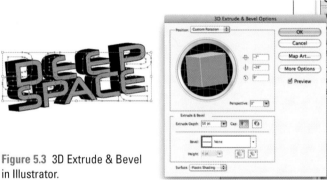

Figure 5.3 3D Extrude & Bevel in Illustrator.

So, with the Illustrator 3D tools you can control an object's angle by rotating the cube. Be sure to click the check box for *Preview. Extrude Depth* is how much depth you want pulled from your outlines. *Bevel* is the shape of the extrusion that is created.

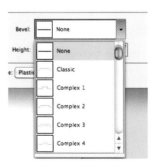

Figure 5.4 Bevel styles.

In addition, you can use *Perspective* to adjust the angle in 3D space that our extruded art sits in.

Figure 5.5 Art with adjusted perspective angle.

Tutorial: A Very Old-School 3D Type Spin Technique Using Illustrator and After Effects

This tutorial is decidedly old school but it has a very cool, lo-fi feel that I can definitely see clients going for in a number of cases.

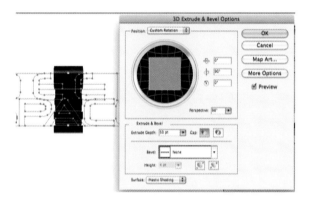

1. In Illustrator, create a type layer and apply *Type>Create Outlines*. Then apply *Effect>3D>Extrude & Bevel*.

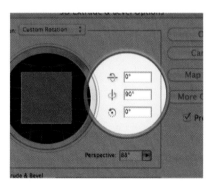

2. In *Extrude & Bevel* set the center axis angle to 90 degrees. For no clear reason, in this lesson I have *Perspective* set to 88 degrees.

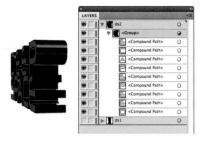 

3. Duplicate our layer. Rename it something sequential (I went with ds2). Open it and highlight *Group*. Go to the *Appearance* panel and click *3D Extrude & Bevel*. Adjust the center axis again, this time to 80 degrees. Notice that the type now turns.

4. Repeat this process as many times as you like, each time reducing the Y axis angle by 10. Save your file.

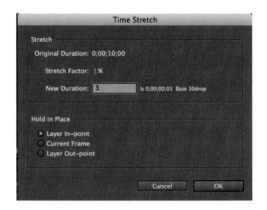

5. Launch After Effects and import the .ai file as a *Composition*. Highlight all the layers and in the *Duration* column set their duration to 3 frames.

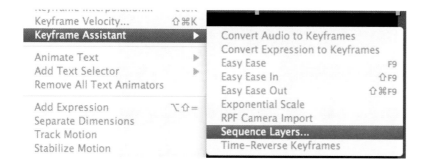

6. Highlight all your layers and go to the *Animation>Keyframe Assistant* menu and choose *Sequence Layers.* It will ask you for a duration of *Overlap*; one frame will do.

7. Press *Preview*, and you'll see your type turning. It's kind of cool. Is there going to be better ways to execute something like this? Of course, and it will be much quicker too. However this is useful, and especially for low-tech kitschy type effects it could work wonderfully. I added our space background and put a *Glow* effect on the type.

Creating a 3D Product Shot

Often motion graphics designers need to create a 3D object that is basically just something that can be done geometrically. If that is the case we can make that happen using a number of After Effects' and Illustrator's existing tools. In this tutorial we'll go through how to set that up.

1. Set your front cover art to one Illustrator layer. Our box will need to have six sides. Let's start by creating another rectangle that is the exact same size as the front cover. We can use that to begin a number of our remaining sides.

2. Copy the new rectangle, until you have three. I set each to a different color from the original cover. I then placed each on its own later.

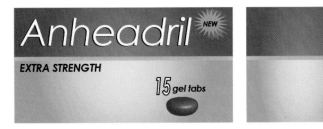

3. Create another layer named rightside. I duplicated the background squares and scaled them so that they were roughly half the size of the other rectangles.

4. This is an optional touch. I made a black x over the rectangles, at a lower opacity, to represent the folds of the package. Although it doesn't really look very realistic right now, it will work when it's used in After Effects. Duplicate this layer and rename it leftside.

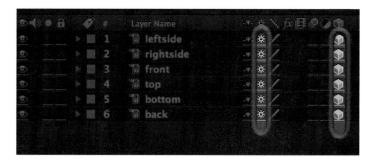

5. Import the Illustrator file in After Effects as *Composition Retain Layer Sizes*. Turn on *Continuous Rasterize* and *3D* for each layer.

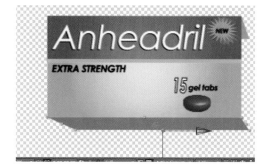

6. Rotate the top and bottom layers so that they are 90 degrees on their X axis. Use the *Selection* tool to maneuver them into their right spots relative to the front layer. The front layer should be brought forward to –60 on its Position Z axis.

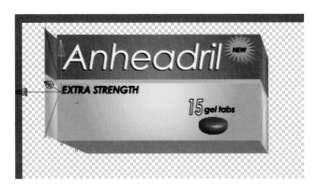

7. Make the leftside and rightside layers 90 degrees on their Y axis rotations and then use the *Selection* tool to place them. The orange line is a great guide.

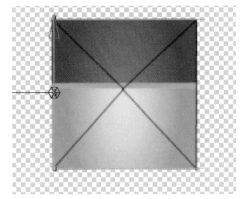

8. The back layer should not need to be rotated. Make it 60 degrees on its Position Z axis, and then use the left or right view to make sure it is positioned correctly.

9. Create a Null Object layer. Rename it Box Controller. Turn on the *3D Switch* and *Continuous Rasterize*. Parent all the layers to the null object.

10. Now, you can animate the Box Controller layer. I made a quick cloudy blue background. Clients will love having a 3D box of their product.

Using Illustrator CS5's New Perspective Tool for a Multipane Shot

In version CS5 of Adobe Illustrator the *Perspective Tool* was introduced after years of Illustrator artists clamoring for it. The *Perspective Tool* essentially creates locking guides for your Illustrator objects in perspective. So in the following tutorial we'll recreate a classic 2D animation technique for creating 3D depth, using a multipane camera. Famously used in the introduction sequence to *Pinocchio*, the multipane camera has different levels for the aspects of the frame to be mounted, creating a sense of depth with the camera's depth of field and positioning of the 2D scene objects.

1. Create a blue gradient background to use for the sky in this scene.

2. Choose the *Perspective Grid* tool in the *View* menu and choose *Perspective Grid>One-Point Perspective>1P Normal View*.

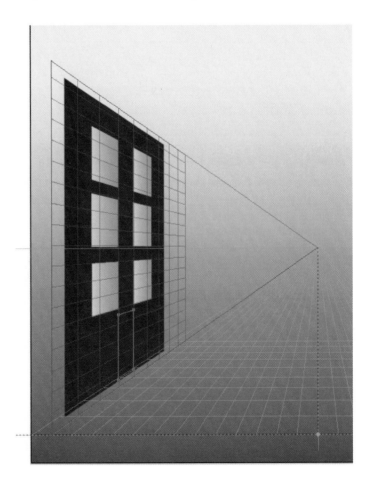

3. Using the *Rectangle Tool* create the shapes; they will snap to the grid. For the windows, create the first rectangle and hold *Option* while using the *Perspective Selection* tool  (under

the *Perspective Grid Tool*) to duplicate it for the next windows. If you use the standard *Selection Tool,* it will ignore the perspective grid.

4. In order to expedite things a bit, duplicate the whole building a number of times, using *Perspective Selection* to keep the duplicate buildings on the perspective plane.

5. Duplicate the whole building row layer and move it to the right side.

6. Complete the scene by adding different elements for a sun, clouds, and a street, each on their own layer. Save the file.

7. Import our scene into After Effects as a *Composition* and turn on all the *3D* and *Continuous Rasterize* switches.

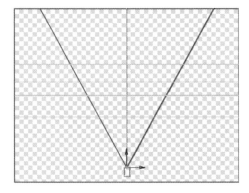

8. Switch to the Top view. Adjust the Z Position of the sky layer to be the farthest from the front, and put the sun and cloud layers in front of it and make a new camera layer.

9. Set a stopwatch for Z Position on the camera layer. Over time have it move backward. Also, turn on *Depth of Field* and animate the *Aperture* to decrease over time.

10. To emphasize the effect, animate the sun's Position from the street to the upper right corner.

11. Experiment with adding to the effect; try duplicating the building row layers and adjusting their positions to make the buildings appear more numerous than they really are.

AFTER EFFECTS 3D CAPABILITIES AND PLUG-INS

The 3D Switch

Figure 6.1 The After Effects 3D switch.

After Effects was not born with 3D capabilities. However, with version 5 in 2001, the 3D layer switch was introduced. The 3D layer switch enables a flat layer to be turned in 3D space. It also makes it possible for *camera* and *light* layers to be used. It is not true 3D; it has become known as 2.5D, and although it doesn't sound that powerful, it is actually quite powerful. Many artists can resolve a large number of 3D concerns by essentially faking the process with 2.5D.

Once the switch is on, flat objects turn in 3D space. So, for example, the *Rotation* tool now has four keyframe parameters, X, Y, and Z rotations, and an *Orientation*. In Figures 6.2 and 6.3 the Y rotation is set at 45 degrees, and the result is that the square has turned in 3D space. After Effects' 3D switch is not an extrusion tool; however, Photoshop has Repousse, which can function as a extrude tool for After Effects.

Figure 6.2

Figure 6.3 A 3D rotation on the Y axis.

Camera Layers

Figure 6.4 Camera layers.

A camera layer in After Effects is a virtual camera, like the ones in big 3D programs. It's incredibly useful in creating scenes that have depth, and the viewers can be given a roller coaster ride through a scene. Camera layers have many controls similar to a real camera. The camera, though, doesn't have a physical appearance in your composition; it can be animated all throughout a scene, with the controls shown in Figure 6.5.

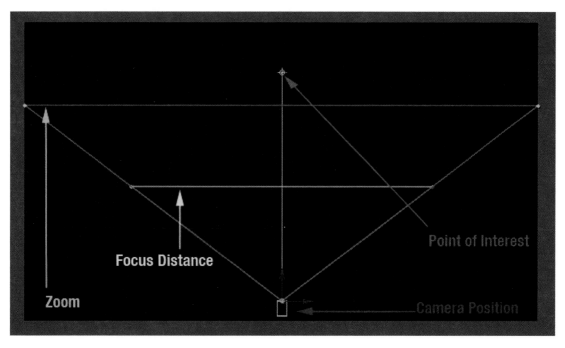

Figure 6.5 After Effects camera controls.

You will have keyframe parameters for *Zoom, Depth of Field, Focus Distance, Aperture*, and *Blur Level*. We will try these out in a tutorial, right after we look at light layers.

Light Layers

Just like with camera layers, light layers only have an effect on 3D layers. Essentially virtual lights, these can be animated throughout scenes just like our cameras. You can choose from four different kinds of lights, *Spot, Parallel, Point*, or *Ambient*. A parallel light would behave like the sun, a spot light is like a stage light constrained by a cone, a point light is meant to be a lower level single light source like a lightbulb, and ambient is meant to affect the light level of an entire scene. In most cases, for a motion graphic spot lights make the most sense, but we should always use the light type that will work the best for the specific kind of scene we are working on.

3D Photos

The following tutorial will demonstrate using 3D layers, cameras, and lights in After Effects. It's become a very famous technique, as it separates a still into separate layers and adds depth between the separate parts. It essentially creates a subtle 3D scene from a still photograph, and can make a piece of interesting footage out of an image that would normally just sit there.

TIP

This technique originates from the 2003 documentary film, *The Kid Stays in the Picture*, which was made largely from stock footage and animated photographs. I highly recommend seeing the film; the technique is used in an excellent way!

1. We begin this project by opening our original photo in Photoshop, and we will break our image into separate parts. Start by selecting the bull, and copy and paste the bull to a new layer.

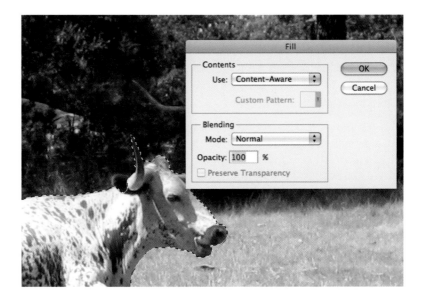

2. Turn off the visibility of the copied bull layer, and with the selection still active, go to *Edit>Fill*, and in the Fill menu, make sure under *Contents* it is using *Content Aware. Content Aware Fill* is a borderline magical tool; for a project like this it makes removals nearly seamless.
3. I did say "nearly" so we'll use the *Spot Healing Brush* to clean up the rest.

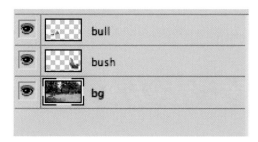

4. I do the same thing with the bush on the other side (removing the bush is a little more challenging than the bull). So now I have a bg layer with the bull and bush removed. Don't worry if *Content Aware Fill* and *Spot Healing* didn't lead to the most perfect removal. We don't need it to look like they were never there.

5. Import the bull scene into After Effects as *Composition Retain Layer Sizes.* Turn on the 3D switch for each layer.

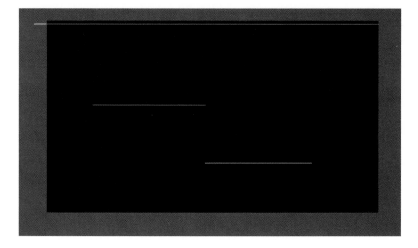

6. Switch to the top camera view and adjust the Z Position of each layer. Think of this composition as set pieces on a stage for a play: push some objects forward and some farther back, and create a sense of depth. I put the background at a Z Position of 350; the bull is at 45, and the bush is at −175.

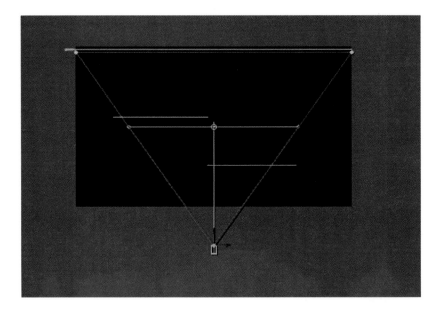

7. Create a new camera layer. Use the *15 mm* preset. Turn on the switch for *Enable Depth of Field*. Press *OK*. In the timeline, on the camera layer open *Camera Options* and set the *Zoom* to 870 pixels.

8. Set keyframes to animate the camera pulling away from the bull. Use the Z Position of the camera.

9. For the complete effect go to *Layer>New>Light*. I chose a *Spot* light. Then, from the top camera angle, I positioned the light and parented it to the *Camera 1* layer.

10. This technique is very valuable and is great for turning the right photograph into a piece of footage. Also, Photoshop CS5's *Content Aware Fill* makes it easier than ever to use this effect on a wide variety of photos.

Importing and Animating 3D Models in After Effects

If you have a 3D element that you'd like to work with in After Effects, you have to take a somewhat roundabout method to bring it into AE. After Effects won't directly take a 3D model, but it will take one in a .psd file as a Photoshop 3D file. Photoshop will work with the following file formats for 3D models: U3D, 3DS, OBJ, DAE, and KMZ. The following tutorial will show you how to take a 3D model to After Effects.

1. Open a 3D model of any of the previously listed formats, and edit it as you see fit. When you are done, save your model as a .psd file.

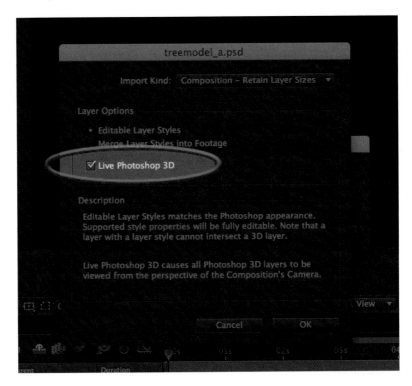

2. Import the .psd file in After Effects. Either *Composition* import type will work, provided that when the Photoshop import options window opens you turn on *Live Photoshop 3D.*

3. Every Photoshop 3D imported layer will come in with three layers, a camera, a null controller layer, and the layer content itself. Everything you need to animate the movement of the object is done on the controller layer. If you want to affect the appearance of the object, then apply that the bottom content layer.

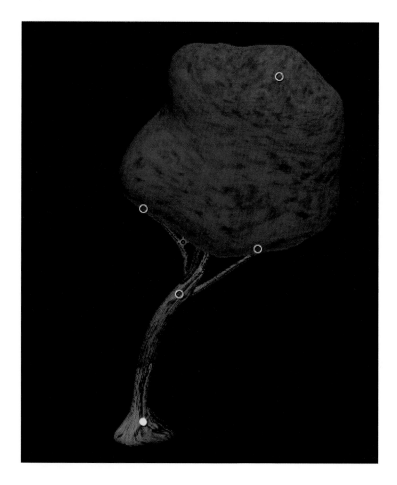

4. To close this tutorial I want to mention a real powerhouse technique. Precompose the composition containing the 3D object. Use the *Puppet* tool on the precomp. Now you have great morphing animation capabilities of the 3D object.

Complete a Sports Logo with a 3D Baseball

Thanks to some bundled effects, some primitive 3D objects are easily added within After Effects, so in the following tutorial we'll use *CC Sphere* to add a 3D baseball to a sports logo.

1. Import the baseball_logo.psd as *Composition Retain Layer Sizes* into After Effects.

2. To add our 3D baseball, import baseballpattern.png as footage and add it to our timeline. I scaled it down to 30%.

3. From *Effect>Perspective* apply *CC Sphere* to our baseball pattern layer.

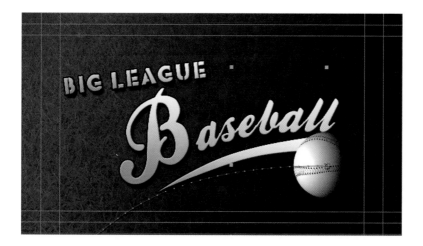

4. Now, let's have some fun with animating our baseball. I created a 1-second-long animation of the baseball coming in from off screen, following along the underline of the word baseball. However without the baseball scaling and rotating this looks pretty lame.

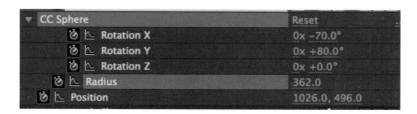

5. Synced with the *Position* animation, under the *CC Sphere* effect, animate the ball's *Rotation X, Y, Z*, giving it some spin, and also include an increase in the *Radius* that will scale our ball. When synced with *Position* it feels as if it is coming toward us. Also, don't forget to use *Motion Blur* on this one.

6. One final point: When creating a motion graphic that contains a very animated element like our baseball, it really helps sell the design if the animated element can have some kind of dialog with the typographic and design elements. So, to achieve that here, I used an animated *Mask* to reveal the underline of the word baseball as the baseball is flying by. It's pretty simple to add this too: Use the *Rectangle* tool on the line layer, and use the *Pen* tool to move the points so that at the beginning of the ball passing by it, the line is completely hidden. Then set a keyframe for *Mask Path* and go to the end of the baseball's animation and move the points so the line is completely revealed.

Organic Lines with Trapcode's Particular

There are few things that Adobe overlooked when they built After Effects, but that hasn't deterred other publishers from making plug-ins that can be added on to the existing After Effects functionality. Over the years some of the most useful ones have become part of the AE package (Foundry's Keylight, the Cycore bundle, Color Finesse, and Digieffects' Freeform, to name a few).

However, if you ask any After Effects artist, the one plug-in that you absolutely cannot live without is Trapcode's Particular.

Particular is a 3D particle engine that provides users with so many options that it is easily one of the most powerful additions available for After Effects. Particle systems are designed for recreating the complex systems that are often found in nature. Basically Particular gives users control of a fountain of particles and their behavior with tons of real-physics parameters.

The Particular effect is so powerful that if I wanted to cover everything that could be done with the tool, it could nearly fill a book. So what I decided to do was to show a popular use of Particular, which is to create animated organic 3D lines.

1. Import the document called organic_lines.psd as a *Composition Retain Layer Sizes*. It only possesses two elements, a background and a 3D-ish cube that we will use as part of the Particular effect. Create a new *solid* layer and rename it *Particular*.

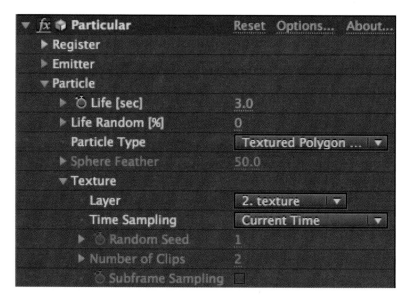

2. Apply *Effect>Trapcode>Particular* to the solid layer. Particular comes with a number of particle shapes built in, but for this project we will supply our own. In the effect controls open the tab for *Particle*. From *Particle Type* choose *Textured Polygon Colorize*. Open the *Texture* tab and choose the layer named texture. Now we are feeding that layer into *Particular* effect. One last thing: To avoid confusion, turn off the eye for the layer "texture"; we only needed it to supply our shape for the effect.

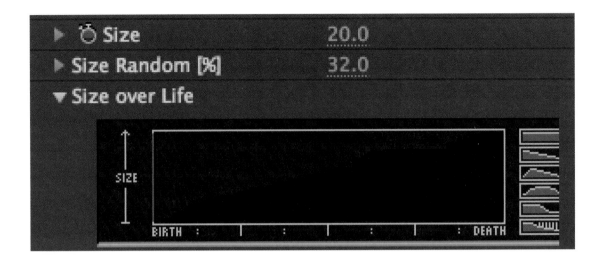

3. While still under the *Particle* tab, head down to size and set it to 20. Also, open the tab for *Size over Life* and you'll see an input scheme that should be new to you if you have used Particular before. Basically this is a graph where you can draw what you'd like for it to do. I drew it so that over the particle's life span the size will increase.

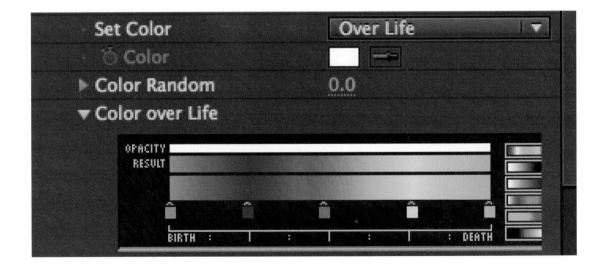

4. Next head down to *Set Color* and change it to *Over Life*. Open the tab for *Color Over Life,* and enter the colors you'd like it to be over its life span.

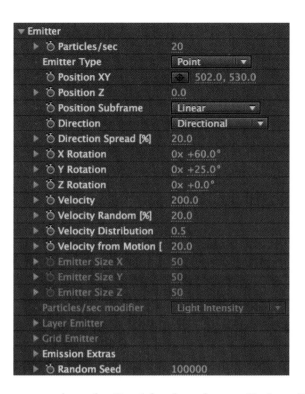

5. Now we can close the *Particle* tab and open *Emitter*. Emitter addresses how the particles are fired from the virtual cannon. Set the *Particles/Sec* to 20, *Direction* to *Directional,* which will allow us to control the flow of particles with the *Rotation* tools. I set the *X Rotation* to 60 degrees and *Y Rotation* to 25 degrees.

6. What you have at this point should look similar to the image above. Under the *Physics* tab, set the *Air Resistance* to 0.1. This is kind of like air pushing the particles back down.

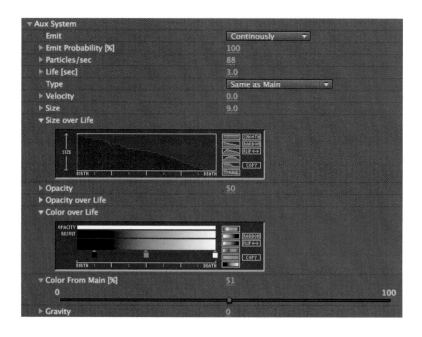

7. Here is where things start to get really cool. Open the tab for *Aux System*. The Aux System is like a duplicator of our original particle flow. Set *Particles/Sec* to 88. You can experiment with the sizes, but I liked the result I got at 9. I set the *Size over Life* on the opposite ramp as the one for our original particles, decreasing over time. I set the *Color over Life* to the preset violet to black but then set the *Color From Main* to 50%, because I want the original colors, but without the 50/50 mix—they just look too bright for me.

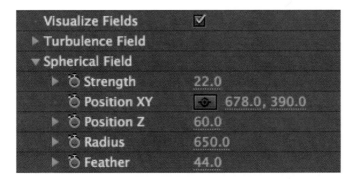

8. Open the *Air* tab and turn on *Visualize Fields*. This will allow you to create fields for shaping the movement of your particles. Open *Spherical Field.* Set the *Strength* to 22 and *Radius* to 650. Experiment with the other settings.

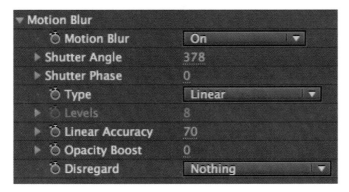

9. Open the *Rendering* tab. Make sure that you enable *Motion Blur.*

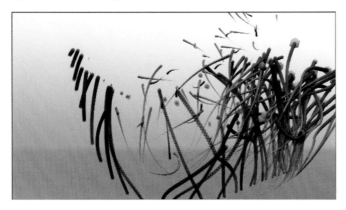

10. To complete this very cool effect try animating the *Particles/Sec* from 20 to 50 and then back to 0 before the end of the timeline. Also, try moving the emitter around the screen. Experiment with this effect because there are so many great things to discover with this powerful tool.

Creating a Laser-Etched Logo Effect with Trapcode's 3D Stroke

Although it's not quite as essential as Particular, Trapcode's 3D Stroke has quite a bit to offer. Just like the packaged Stroke and Vegas effects, 3D Stroke is applied to a layer with a path; the difference is that 3D Stroke is a 3D effect. In addition it has that excellent Trapcode touch. The following tutorial demonstrates what it can do.

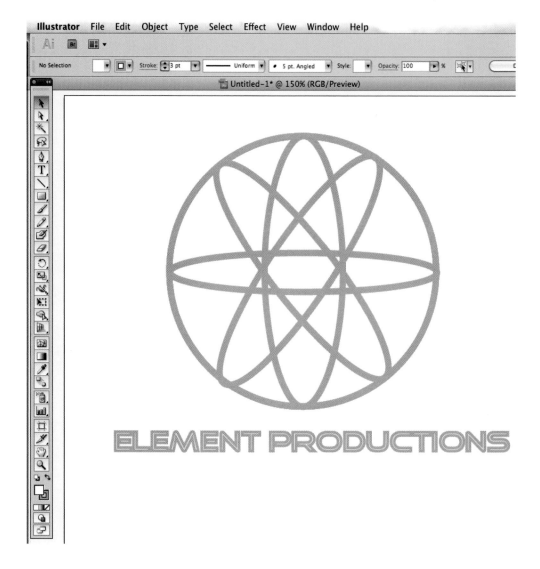

1. Design your logo in Illustrator. Though we will only use the paths from this design, they are easier handled in Illustrator or Photoshop than in AE.

2. Import the bg.psd into After Effects and then create a new *Solid* layer. In Illustrator highlight the atom-like element and press Ctrl/Cmd-C to copy and Ctrl/Cmd-V to paste it onto our solid layer. You should now have a corresponding *Mask* for each path from Illustrator. I then renamed the solid Laser Logo.

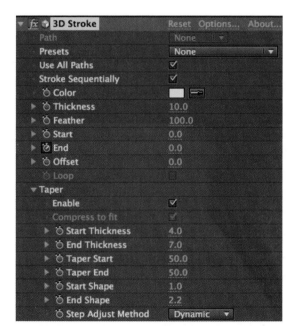

3. Next apply *Effect>Trapcode>3D Stroke.* Turn on *Use All Paths,* which applies the effect to every mask, and *Stroke Sequentially* so the *Start* and *End* functions will follow the paths in numerical order. Under *Color* I chose a yellow. I set the *Thickness* to 10 and *Feather* to 100. At the beginning of the timeline, I started a stopwatch for *End* and set it to 0. At the end of the timeline I set another keyframe for *End* to 100. This will animate the stroke traveling around our sequential paths from start to finish. Next, open up *Taper* and *Enable* it. Set the *Start Thickness* to 4 and the *End Thickness* to 7.

4. To give the logo some much needed slickness and flare, apply Trapcode's *Starglow*. It will give it a nice laser feel and some color dynamic. Set the *Preset* to *Tri Prism* to get the colors I am using, but also feel free to experiment with other looks. I set the *Streak Length* to 15 and the *Boost Light* to 2.

5. The next thing I wanted to do was to have the camera follow the tip of the beam as the logo is drawn. So inside *3D Stroke*, scroll down until you see *Camera* and open that tab. Start stopwatches for *XY Position, Z Position,* and *Zoom.* Adjust these to keep the front of the beam in the frame the whole time. To have it gradually reveal the whole logo, over the course of the timeline have the *Z Position* go from –35 to –130.

6. To complete the animated graphic I added in the type and used the preset *3D Fall Back Scale and Skew.*

Interview: Mayet Bell

Mayet Bell is a 3D and 2D artist. She earned her MFA in Film and Animation from The Academy of Art University in San Francisco, and her BFA in 2D Animation and Illustration from The University of the Arts in Philadelphia.

Mayet's work experience is in the games industry, where she worked as a menu and props artist at 2K Sports in Novato, California. She modeled, textured, and animated interactive menus using 3D software. She also worked on props modeling and texture work on player faces, trophies, and stadiums. Other games experience includes contract work for Zynga Games. While at Zynga she created art assets for eight MMO games on Facebook and Myspace, which included Space Wars, Football Wars, Fashion Wars, Dope Wars, Bumper Stickers, and Heroes vs. Villains.

Along with working in the games industry, Mayet has worked as a freelance illustrator, designer, and 3D modeler for commercials, independent film, and individual clients. She has also launched a line of T-shirts online for both men and women.

Figure 6.6 Mayet Bell, 3D and 2D animator and illustrator.

Mayet is currently a full-time faculty member at the Art Institute of Austin, where she teaches 3D Animation, Rigging, Final Project, Background Layout and Design, and Production Team.

What compelled you to start a career in 3D animation?

I had always done art before, and then I saw *The Little Mermaid* and I was like "Oh! That's what I want to do!" So I went to undergrad for 2D Animation, with a minor in Illustration, and that is when, about 1999, the 2D animation field imploded. I had always liked 3D, I always wanted to learn 3D. I took one class in SoftImage as an undergrad and that was it. I realized that if I really wanted to be hireable, I needed to learn 3D, so I went back to school. I got my Masters in 3D Modeling and Texturing, and I found it very helpful that I had a 2D background, going into 3D.

Continued

Would you say that being trained as an illustrator has made it easier to learn modeling?

I think it is helpful when you are creating your own characters when you are concepting out stuff, you can make the image planes. It helps you to be able to model your own characters and not to have to seek out someone else's. So I think that drawing is very important, also in terms of proportion. If you have good drawing skills, you tend to be better at creating correct proportions in a 3D space.

When you first started doing 3D animation, what was the first software package you used and why?

SoftImage was the first one I learned because that was what they had at my undergrad. And when I went to grad school Maya is what they taught us, but because I learned SoftImage, even though it had been years, I knew the basics and I picked up Maya faster.

Can you tell me what your workflow is on a project?

If it is an environment I'll typically find reference, or if it's drawn I'll get my concept design and use that for reference; typically I'll import the reference on a plane, and import it into my scene just so I can see it. And then I usually block out everything, so I'll block out, I'll just do the simple shape of the model and model as much as I can until I can't go any further without upres-ing the model. Then I go from there.

What's your favorite movie?

For science fiction, *The Empire Strikes Back*; for drama, *Elizabeth I*. For animation it's a toss-up right now between *Howl's Moving Castle* and *The Incredibles*. For romantic comedy, I like *The Proposal* right now and *Singing in the Rain*. It tends to change.

Can you share a moment when you found something really inspiring?

Probably *Elizabeth I*; that was the first time I saw experimental techniques used in film in a good way, used to illustrate how the character is feeling. When Elizabeth finds out she is going to be queen, when her sister Mary has died, everything is kind of a blur; initially we see her and her ladies in waiting through some glass and it's all very blurred. When she goes outside to accept the ring the film becomes overexposed. While she is accepting the ring, it gets overexposed again. I thought it was a really great way of showing the daze that she's in, because she almost died, and now she is going to be queen. The concept is matching the technique. It was really cool and all those horrible experimental films I had to watch make sense now, it's being used well.

What is one skill that every 3D artist needs to understand and master as soon as possible?

I'd say attention to detail. Whether it is proportion, texturing, lighting, that all comes down to details. Even for 2D artists it is very important. What is this environment trying to tell the audience? What is this character, how they look, trying to tell the audience? Just from looking at the character. I think attention to detail is the most important thing.

Where do you look for inspiration?

Spectrum. I really like the *Spectrum* series of Illustration. I go to CG Society; they have really beautiful 3D art there. I get the *3D World* magazine; I surf people's blogs, art blogs; any art book that I see that is awesome, I am like "*must buy!*"

What is your dream project?

If ILM called up and was like, "Hey Mayet, Want to work on anything?" I would be like, "Yes, Please!" I could not turn that down. Or Disney, if I could work on a Disney or Pixar animated film. I don't know if that will ever happen, but that would be fun.

What have you done so far that you are most proud of?

I tend to get bored with the projects I've finished very quickly. So, I guess in 3D this dalek I've recently finished. In my 2D work, my self-portrait, the happy version of myself.

Figure 6.7

What skills do you feel like you are still learning or improving upon?

Lighting has always been a weak point for me in 3D. I even had a hard time grasping lighting in 2D. Color is my strength. I got color right away but lighting for me is something I'm always working on.

What have seen that you wish you worked on?

Paprika is a really awesome 2D/3D animated film that would have been cool to work on. The art on the trade paperback *Catwoman*, that is sickly awesome. I would have loved to have been somehow involved in that. In movies, I really loved the visual effects for *Captain America*.

BEGINNING 3D FOR REAL

Modeling

In previous chapters we found many ways to create 3D elements in After Effects, so why would we need to learn about the complex ways of dedicated 3D software? In fact, when it comes to adding the 3D skillset to your 2D skillset the thing that most people really need is the knowledge of creating 3D models. Keyframe animation is largely the same no matter which software package you use. After Effects doesn't allow you the full range of ability to create 3D models that a dedicated 3D software package can. So in this chapter we will look at the process of creating 3D models in 3D software.

3D models combine two familiar approaches for creating computer graphics. A 3D object is a shape determined by a vector mesh—this means that the mesh could be resized without quality loss. Much of 3D modeling is done by adjusting vertices in a similar way as drawn objects are adjusted in Illustrator. Applying a raster image to the exterior of the 3D mesh creates the surface appearance of a 3D object. So, the raster graphic sits on top of the mesh, and if the mesh is complex enough and matches the exterior image well, the 3D object will take on a photorealistic appearance.

3D models begin by establishing the correct geometry. Adding a surface should be done later. It's also important when creating a 3D model that you find the appropriate *reference*. References can be photos or videos, but you need materials to look at to make sure you are building the thing you want to in a believable way. It's important to have references even if your plan is to make something very stylized, or unrealistic. With stylized or unrealistic models there still needs to be touch points from our reality in order for the object to be believable and to read well.

Getting Started with Cinema 4D

Figure 7.1 Cinema 4D's screen.

Views

As mentioned earlier in the book, I chose to focus on Cinema 4D because of how well it works with After Effects and because it has a relatively quick learning curve. Figure 7.1 shows you the main screen of Cinema 4D. Most 3D applications will have the ability to swap between multiple views. C4D is no exception. Typically you'll want to swap between having the *Perspective* view up and having all four views. You can assign a different view to each port if you prefer.

Figure 7.2 Viewport tools: move, zoom, rotate, and toggle active viewport.

When modeling, it is vital that you move around your 3D object often to see what is happening. You can't ever do everything from one view (remember, it's 3D, so things may look right on one side but could be completely off elsewhere). Each viewport has camera controls, as shown in Figure 7.2, to move, zoom, and rotate your way through the current view or to swap between views.

Tools

Figure 7.3 Tools: *Undo, Redo, Select, Move, Scale,* and *Rotate.*

These tools should feel very familiar: *Undo, Redo, Move, Select, Scale,* and *Rotate.* On the *Select* tool you will see an arrow in the corner, which means the same as it does in Adobe software—that there are more tools underneath.

Figure 7.4 As you would expect, there's more than one way to make a selection.

Moving to the right, you'll see three more groups of tools. First is the tool by itself, which is just a history of tool selection; if you hold it down you'll see the last few tools selected. Next is the group of axis-locking tools, and then the last three are rendering tools. We'll be revisiting this area later.

Figure 7.5 Tool history, axis tools, and rendering tools.

Now once again moving to the right is our group of object tools. In Cinema 4D any element of a 3D scene is an object. Here you can find objects and their modifiers.

Figure 7.6 Object tools: *Primitive, Spline, NURBS, Modeling, Deformation, Environment, Camera,* and *Light* (moving left to right)

Now, we are going to get started with making our first model. As we begin this, pardon me for a few brief thoughts on the *philosophy* of modeling. Essentially, we have a set series of techniques that we use to make things in 3D. Most objects in our world are made from combining and altering different shapes. Whenever I plan to model something, I look at the objects and try and figure out the best method to get there. My reference is a screwdriver.

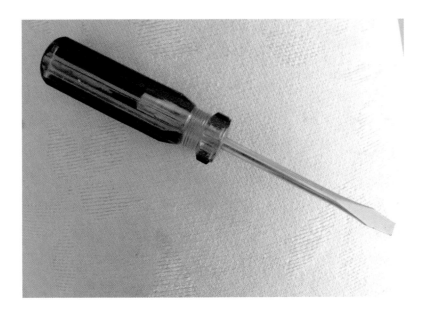

Figure 7.7

Look at the object and think about how we can get there. Think about our set of primitive shapes. I imagined what is shown in Figure 7.8.

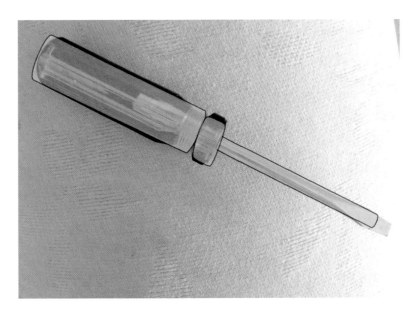

Figure 7.8 Screwdriver in primitives.

Looking at my reference, I forget the screwdriver shape for a moment and think about the shapes that make it up: cylinder, ring, cylinder. Is this the exact way we'll get there? Maybe, maybe not. But 3D modeling trains the human mind to deconstruct objects into combinations of warped primitive shapes.

Basic Modeling from a Photograph

1. Let's begin by importing our reference photo into Cinema 4D. Toggle your viewports so that you see the four-viewport view. Click the *Top* viewport. From the viewport choose *Options>Configure*. This will open the viewport's *Attribute Manager*. The Attribute Manager usually lives in the lower right-hand corner, and it controls aspects of how any element in C4D is handled.

2. In the *Attribute Manager* choose *Back*. You'll see *Image*; then click the ellipsis button to the right. Choose our reference image.

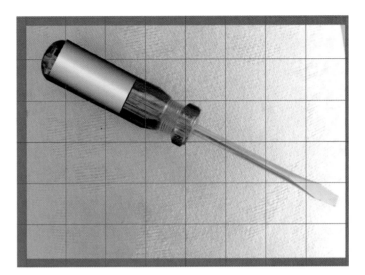

3. Maximize the *top* viewport. You will see our reference image in the background. Click and hold the *Add Primitive Object* button and choose the *cylinder*. Use the *Scale* and *Rotate* tools until you have adjusted its place to match our object.

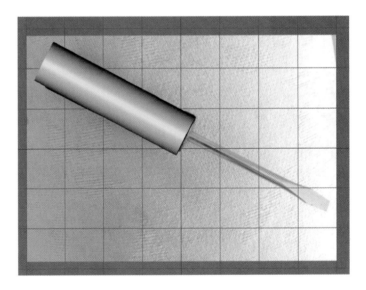

4. Our ability to edit this object will be limited until we click on this icon or press the letter C on the keyboard. This makes the vertices on our cylinder object editable. Turn on the *Model Tool* and now you can adjust the scale without the size ratio locked, and you can also edit the vertices.

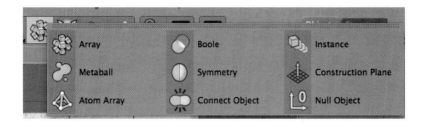

5. Go to the *Modeling Tool* menu and create a new *Boole* object. A Boole is an object that uses one shape to remove overlapping area from another shape.

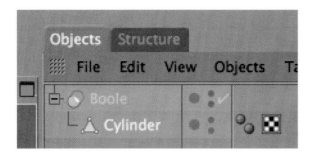

6. In the Object Manager, move our cylinder inside the Boole object.

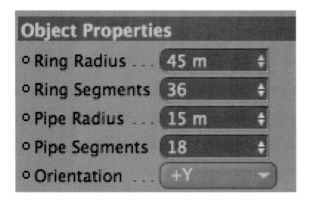

7. Create a new *Torus* object, and in the *Attribute Manager* under *Object Properties*, set its *Ring Radius* to 45 and *Pipe Radius* to 15. Then position the torus over the part of the screwdriver's handle that dips in.

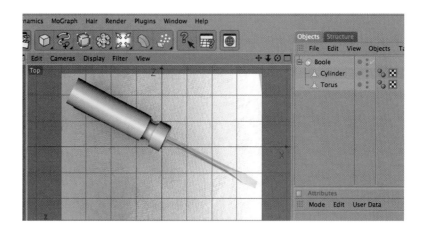

8. Place the torus below the cylinder inside the Boole object. Now you'll see the torus shape cut from the screwdriver handle.

9. Highlight the Boole, torus, and cylinder. Inside the *Object Manager* go to the *Objects* menu and choose *Connect + Delete* (or *Connect Objects + Delete*, depending on your version of C4D). This will merge everything down to a single object and then delete the three parts that made it.

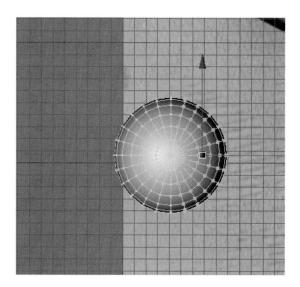

10. Let's finish up this handle. I want to have a round cap at the top, so I'll make a new sphere object. Press the C key to make the object editable. Now, from my select tools, I choose *Rectangle Selection* and highlight half the points. Press Delete. If there are more points left, repeat the process until you have a half-moon shape.

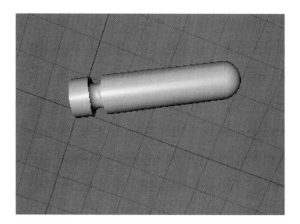

11. Line up our half-moon shape until it fits nicely at the top of the screwdriver handle. Check how it is lining up from multiple viewports, and when you are happy, go ahead and choose *Connect + Delete* from the *Objects* menu in the *Object Manager*.

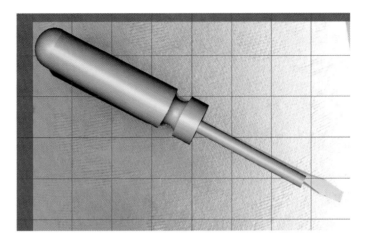

12. Return to the top view. We now need to create the rest of the screwdriver. Create a new cylinder, make it editable, and scale it to match the screwdriver. Don't worry about the tip at the moment.

Changing Views

This now brings me to a point I wanted to make about how important it is to change your view constantly. While working on step 12, you may have something like this:

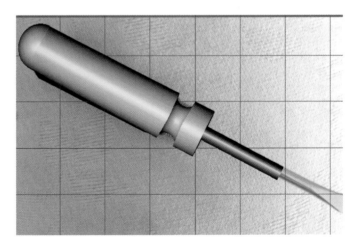

However when designing in 3D space, remember, unless you change your viewing angle often, you are not seeing the whole picture. Look at what happened when I changed my viewport.

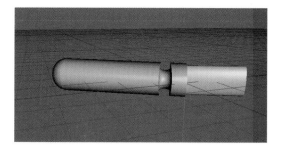

That cylinder is not nearly as round as I thought. It's easily fixed, but I just wanted to point out how things may not look the same from other views.

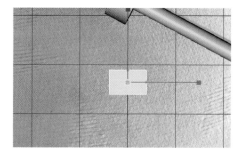

13. Let's start making our tip for our screwdriver. Make a new cube, and make it editable, and scale it to about the same size as the tip of the screwdriver.

Figure 7.9 Editing tools: *Point, Edge,* and *Polygon.*

When modeling there are different approaches to how you may want to edit your shapes. The *Point Tool* (the top button in Figure 7.9) is for editing individual vertices. The *Edge Tool* will allow you to edit the edge of a polygon, essentially any two points. The *Polygon Tool* will edit an entire shape, or four points (for the sake of simplicity, I will be sticking to quads for this book, and avoiding tris).

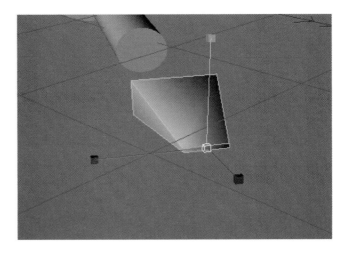

14. Using the *Polygon Tool*, I scale only the front of the tip down to get the shape that looks like our flathead screwdriver.

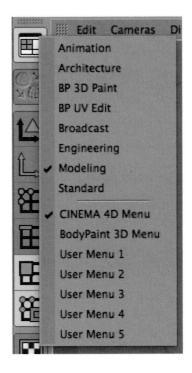

15. Cinema 4D has preset layouts in the upper left corner or upper right depending on the version of C4D you own. Open that menu and choose *Modeling*. Now you'll have access to a bunch of great modeling tools.

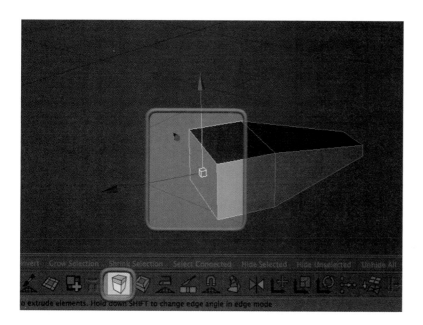

16. Select the polygon at the rear of the tip of the screwdriver. From the Modeling tools at the bottom choose *Extrude Selected*. This will give us a new polygon that we can move. I drag mine farther back as shown in the above figure.

    A great deal of time modeling will be spent creating a set of vertices or points and then *extruding* those points. An extrusion simply means a pulling out, or taking a shape and moving it in 3D space to develop a more complex object.

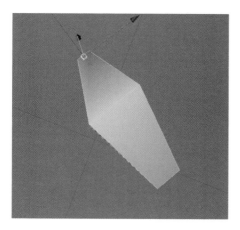

17. Let's scale down that polygon. Looks like my tip is ready. Once it is correctly positioned, merge it with the driver's cylinder by selecting our tip and that cylinder and using *Connect + Delete* from the *Object Manager*.

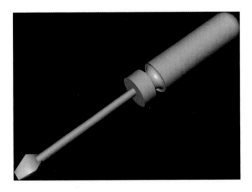

18. You have your first model. To see what it is really going to look like you'll need to render a frame. Press Ctrl-Cmd-R to render the active view. Now, it's all gray, so next we'll be looking at adding materials.

Basic Materials

A 3D material is a 2D image that will sit on top of the 3D mesh and help create the final appearance of the model. Typically you'll source your materials from a combination of photos and painted images. Both have benefits and drawbacks. However, it's important to understand how both a good model and a good material are vital to creating the best, most photorealistic 3D models. One can't save the other—a bad material on a good model is just as bad as a good material on a bad model.

Introduction to Materials in Cinema 4D

1. With the screwdriver model open from the previous tutorial, on the lower left-hand corner of the C4D screen, find the *Materials Manager*, open *Create*, and choose *New Material*.

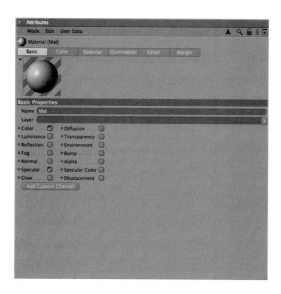

2. You will now have a new material to work with. Let's start with the *Basic* subheading. This is where we can set up our general options for our material. First, rename the material "screwdriver handle" under *Basic Properties*. Next, you'll have two columns of check boxes. We'll go through these one at a time in the materials chapter, but for right now basically these will turn on and turn off different channels that are mixed to create the final material. You'll see a subheading next to *Basic* if that channel is on.

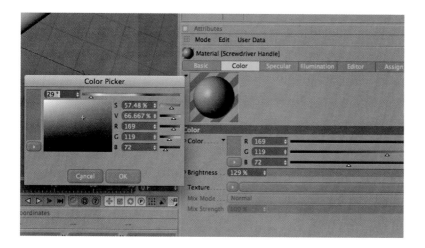

3. Choose the color subheading. Clicking the color swatch will open the *Color Picker*. I chose a dark orange.

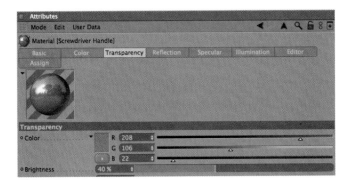

4. Return to the *Basic* tab and turn on *Transparency* and *Reflection*.
    Go to the *Transparency* tab and set the color to an orange, and
    lower the *Brightness* to about 40%.

5. Go to the *Reflection* subheading, and lower the *Brightness* of
    the *Reflection* channel to 40%.

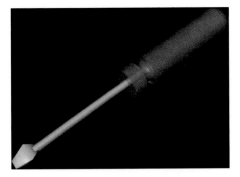

6. Drag and drop the material onto the screwdriver handle. Now,
    press Cmd-R to see a quick render. You'll see the reflective and
    transparent properties in action.

7. Make another new material, name it "driver," turn on *Reflection*, and go to the *Specular* tab. The specular settings define the highlights. Choose the *Metal* mode. I put the *Width* at 90%, *Height* at 80%, and *Falloff* at 40%.

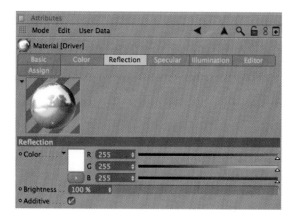

8. Under *Reflection* turn on *Additive*. Now apply the driver material to the model.

9. Press Cmd-R and take a look at how we've progressed here. The right materials really make a model look much more realistic.

Introducing 3ds Max

Though every piece of software has its differences, the core of modeling is based on the same basic functionalities. You can either manipulate primitive geometry or extrude polygons by adjusting points, polygons (or faces), or edges. The following tutorial will introduce you to the 3ds Max interface, and you will see some major similarities to Cinema 4D.

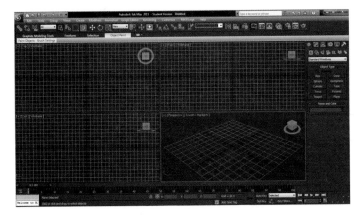

Figure 7.10 The main screen of 3ds Max.

3ds Max starts you off with a four-viewport setup. You can easily maximize a view by pressing Alt-W. In the upper right corner of every viewport is a Viewcube, which is an easy UI device for quickly adjusting your camera angle.

Figure 7.11 The 3ds Viewcube.

At the top of the screen, you have the main toolbar, which has the most common tools you will use to interact with 3ds.

Figure 7.12 The main toolbar.

The first three tools starting from the left are the *Linking* tools designed for controlling how two separate 3D objects should either be attached or detached from one another.

Figure 7.13 Linking tools: *Link, Unlink,* and *Bind to Space Warp.*

The next four tools, moving to the right, are the *Selection* tools designed for selecting either objects or parts of objects. Note that the *Rectangular Selection* has a white arrow in the lower right corner, meaning that there are more tools under it.

Figure 7.14 Selection tools: *Select Object, Select By Name, Rectangular Selection*, and *Window/Crossing.*

Figure 7.15 Standard transform tools: *Select and Move, Select and Rotate, Select and Scale, Reference Coordinate System Dropdown*, and *Use Pivot Point Center.*

The following four tools to the right are our familiar *Transform* tools. To the right of those are two toggles, one for *Select and Manipulate*, which allows you to select and move objects, and the other for *Override Keyboard Shortcuts.*

Figure 7.16 Toggles for *Select and Manipulate* and *Override Keyboard Shortcuts.*

Figure 7.17 *Snapping* toggles.

The next four are *Snapping Toggles*, which give you control over how an object will jump to the grid. To the right of these tools are the *Selection Set* tools, which are designed to let you store a certain selection of points and recall that selection at any time.

Figure 7.18 *Selection Set* tools.

Figure 7.19 *Mirror* and *Align*, *Manage Layer Sets*, and *View* tools.

The next two are *Mirror* and *Align*, followed by *Manage Layer Sets*. To the right of that are toggles for specific views. First are the *Graphite Modeling Tools*, which are tabbed menus just below the main toolbar. Then there are the toggles for the *Curve Editor* and *Schematic View*.

Figure 7.20 Render tools.

The final group of three are the render tools. In addition to the main toolbar, on the right-hand side of the viewport you will also have the *panels*. Panels have *even more* tools for specific purposes. We'll discuss some of these as we model in 3ds.

Figure 7.21 Panels: Create, Modify, Hierarchy, Motion, Display, and Utilities.

Modeling a Stop Sign in 3ds Max

1. Select the *Create* panel, then choose *Shapes*, and below *Shapes*
   choose *Line*.

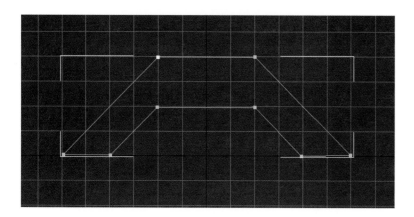

2. Using the *Line* tool, switch to the top view and make a shape
   like the one in the above figure.

3. Go back to your panels, and this time choose the *Modify* panel. Right below the name of your object is a drop-down that is beloved by some and despised by others, called the *Modifier* stack. A laundry list of tools for modifying objects, it is both daunting and powerful. Scroll down to *Extrude*.

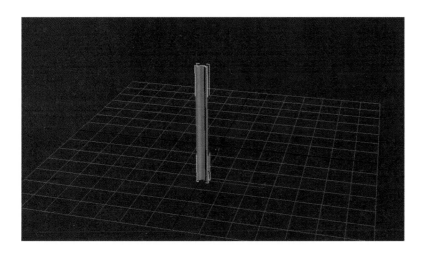

4. Scrub the number value for *amount* until you like the height of the signpost. You may need to scale the object to keep it an appropriate size.

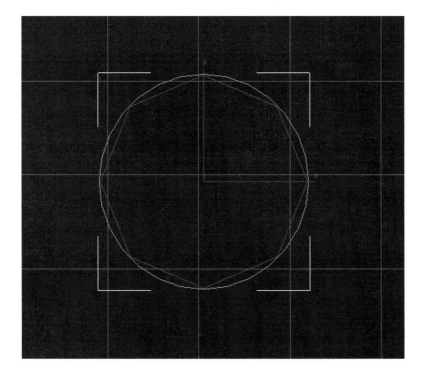

5. Return to *Shapes* and this time choose *NGon*. Set the sides to 8, and then create your shape. Use *Select and Rotate* to turn it on the Z axis (in the top view) so its angle matches a stop sign.

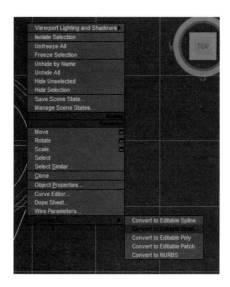

6. Right-click our NGon, choose *Convert To*, and then choose *Convert to Editable Mesh*.

7. When you know the exact amount of rotation you want, it is often easier to type it at the bottom. On the Y axis type 90 degrees.

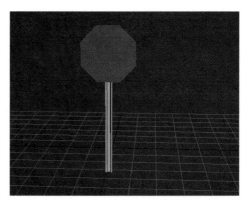

8. Switch back to the Perspective view. Line up your sign to the signpost. Adjust the scale where you feel it is necessary.

9. Now we'll finish this model with some materials. Open the *Compact Material Editor*. The new 2011 version of 3ds Max introduced a more complicated *Slate Material Editor,* which features a nodal interface, but we'll keep it simple for right now.

10. Go to the slot right next to *Diffuse* and click, which opens the *Material/Map Browser.* Choose *Bitmap* and click *OK.* Load the stop sign image and then apply the material to the sign by dragging and dropping.

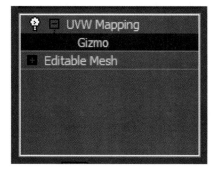

11. Apply the *UVW Mapping* modifier to our sign object. Below, find the heading for *Mapping* and choose *Planar.* Back at the

top expand the *UVW Mapping* modifier and highlight *Gizmo*. The *Gizmo* allows you to use the *Move, Scale,* and *Rotate* tools to position your material.

12. Press the render button and take a look.

Making a 2D Logo 3D in Cinema 4D

In most cases when you receive a logo that will need to be redone in 3D, it will come in the form of an Adobe Illustrator file. No need for alarm—it's a great thing. We can use the paths from the Illustrator file as geometry. In the following tutorial we'll go through how to make a 3D logo from a 2D logo.

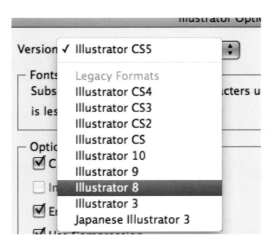

1. No matter how old of a copy of Illustrator your client may have, it's probably more recent than version 8. However, C4D needs the file to be saved in the version 8 format to work with it. So choose *File>Save As*, name the file, and then on the next screen you'll have an option to choose a version; choose version 8.

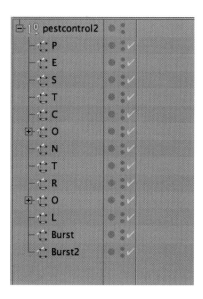

2. In Cinema 4D go to *File>Open* and choose the Illustrator file we just saved. Now, you'll just have paths, and in the *Object Manager* you will have paths that are unnamed. Go through and rename your paths. Some may be grouped in a way you don't like, so expand and rename the paths. On the letter O's you'll see that those are expandable; that is because it takes two paths to make a letter O. Highlight the two paths on the O's and go to *Object Manager>Object>Connect + Delete* to make it so that they are single objects.

3. I highlighted P, E, S, T's paths and in the *Object Manager* under *Object* I chose *Connect + Delete* to make them one object. From

the toolbar, under the NURBS objects, choose *Extrude NURBS* and place our "Pest" object inside of it. Now you'll see the word Pest is extruded.

4. Repeat the process for the word Control and the two bursts. Spread the different objects out in Z space.

5. To bevel the edges of our words, for some stylistic flair, I highlight the Extrude NURBS object, and in the *Attribute Manager* I select *Caps*. Then set the *Start* to *Fillet Cap* and the *Radius* to 3.

6. Experiment with different materials at different settings. On mine I made use of *Glow* on the type, with metallic settings on *Specular*. With the two bursts, I used a lower transparency on the one in front.

ANIMATION IN 3D

Keyframe Animation

Just like After Effects and other 2D animation programs, 3D animation follows the same basic principles of keyframe animation. Set a keyframe at the start point and another end of the motion, and the computer will do the *tweening* (creating frames in between two keyframes). Of course there will be things that we'll have to address that are going to be specific to the software, but many of the general concepts will be quite familiar to you.

Animating the Camera in Cinema 4D

To briefly demonstrate the animation tools in Cinema 4D we'll start with animating the camera. So, let's open our logo from Chapter 7.

1. When starting an animation project, you should visit the *Project Settings*, which can be found under the *Edit* menu. Here you can set the frame rate; I've set mine here to be 30 fps.

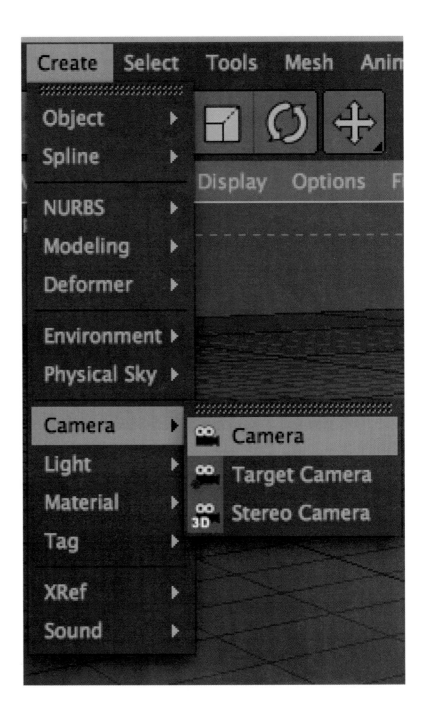

2. From *Create>Camera* choose *Camera*; this will create a new *Camera* object.

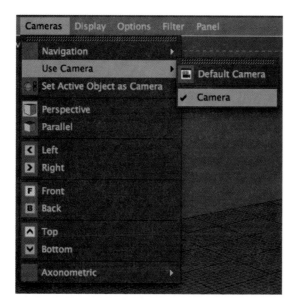

3. At the top of viewport, go to the *Cameras* menu and choose *Use Camera>Camera*; now we are looking through the camera object we just created.

4. Using the viewport controls set your camera angle how you'd like the first frame to look. I zoomed back and went to the left. Then press the *Record* button (it's the key icon below the viewport).

TIP

Remember, in keyframe animation you need at least two keyframes to make something move.

5. Move the playhead (it's the bright green box in the timeline) to the 50-frame mark. Using the viewport controls, zoom in and change the angle to the right-hand side. Now, press the *Record* button again to get your second keyframe.

6. Click the *Play Forward* button to see the movement.

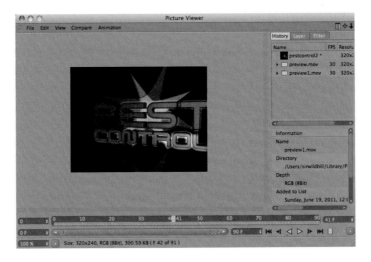

7. To see a more accurate preview with the rendered model, go to *Render>Make Preview.*

Bouncing Ball

A bouncing ball is a great introductory exercise in animation. It exposes new animators to a number of aspects of the tools in Cinema 4D (C4D) designed to make the animation process as painless as possible.

1. For this exercise, it is best if we switch our C4D layout to *Animation.* The Animation layout gives us a smaller viewport but a much larger timeline. It's pretty comfortable for those of us who are used to the After Effects layout.

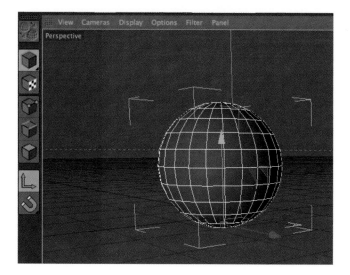

2. Create a new sphere object. Make it an editable object by pressing C on the keyboard. Use the *Object Axis Tool* to move the pivot point to the bottom of the sphere.

Figure 8.1

In Figure 8.1 you'll see our timeline controls. From left to right, we will first see the arrows for *Go to Previous Keyframe, Go to Animation Start, Go to Previous Frame, Play in Reverse, Play Forward, Go to Next Frame, Go to End of Animation,* and *Go to Next Keyframe.*

The next button is for how you want to see your playback. *Cycle* (a loop) is pictured in Figure 8.1 (the other options are *Simple,* which will just play your timeline back once, or *Ping-Pong,* which will play it forward, then in reverse). I never take mine off of *Cycle.*

The next group to the right is our keyframe controls. The key icon records the *Position, Scale, Rotation,* or *PLA* of our active objects. PLA is *Point Level Animation,* which refers to positions of all points on a polygonal object. After the key button is the toggle for *Automatic Keyframing,* which will make our C4D timeline behave more like After Effects where you set the first keyframe. Then every time you are on a frame that doesn't have a keyframe already and you make a change, a new keyframe is created automatically. The question button is for *Set Selection Object for Keyframing* to record the actions of an object within a selection object.

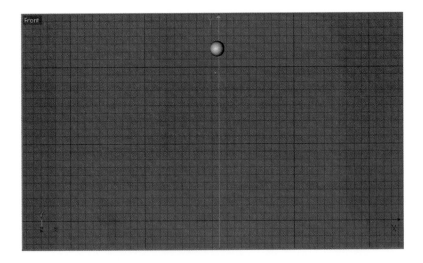

3. Switch to the front viewport. Turn on *Automatic Keyframing.* Move the sphere to the top of the viewport and set your first keyframe.

4. Go to the tenth frame and drag the sphere down to the red line. You will now have another keyframe in the timeline and you'll see a *Motion Path* in your viewport.

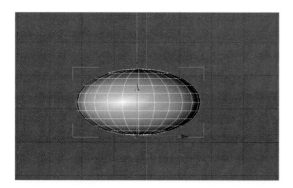

5. Go forward one frame to frame 11. Select the *Object Tool* and your sphere and use the *Scale* tool to stretch out the sides on the X axis. You'll see your ball expand.

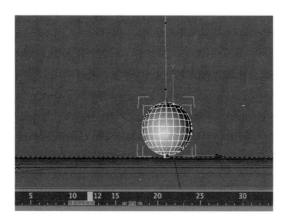

TIP

 By stretching the side of the ball in this tutorial, we are using the important animation principle Squash and Stretch, which basically represents real-world physics in animation by adjusting the size and proportion of an object to make it feel more real.

In C4D you must use the Object Tool in order to animate Scale.

6. On frame 12 return your sphere to its correct proportion. To be certain you return it to the correct size, open the *Coordinate*

Manager from the window menu and type the correct values. I used 200 m originally, so that is what I will return it to.

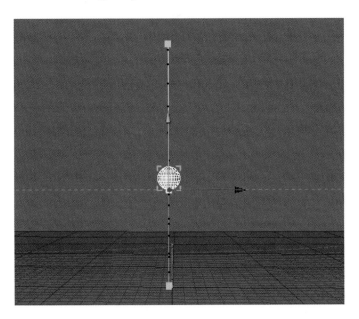

7. Go to the nineteenth frame, and move the sphere up about halfway to where we started it from.

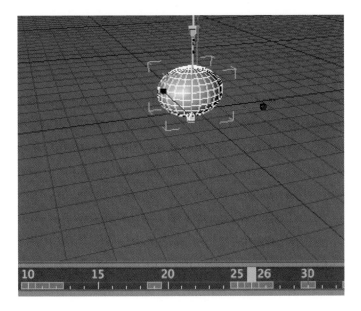

8. At frame 25, drop the sphere back down to the floor again. Repeat the squash move from steps 5 and 6, only this time, less so, as the velocity of the ball is decreasing.

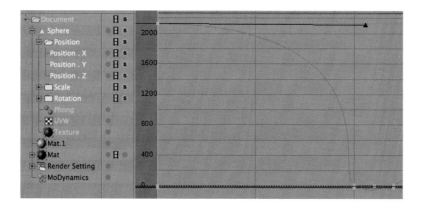

9. In the timeline, under *View*, choose *F-Curve* mode. After Effects users who are familiar with *Easy Ease* will no doubt recognize this. What we see here is the curve of how frames are generated between keyframes. On each point, which represents a keyframe, we have handlebars. If you extend the handlebars as shown in the above figure, the ball will start slower and pick up speed as it approaches the ground. Try adjusting the F-Curve for each keyframe.

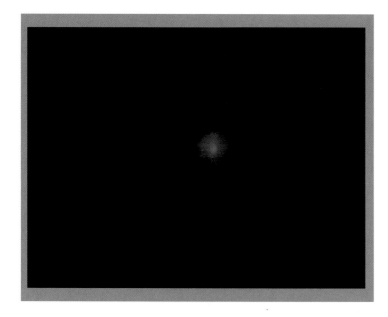

TIP

Slow In and Slow Out is another principle of animation, which states when an object in motion begins its movement it will take some time to accelerate and then slow down again as it decelerates. The speed of the object should not be consistent the entire time it is in motion, unless the point of the animation is to make the object appear robotic.

10. Repeat the drops until it feels like enough bounces. I added a red material to make it feel more natural. Also, when I previewed the file, I added some motion blur by going to the *Render Settings>Effect*.

The Bouncing Ball in 3ds Max

Now, we'll go through the same tutorial with 3ds Max.

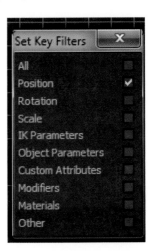

1. Create a new sphere, and maximize the front viewport. If your model is in wireframe it might be easier to work with if it is *Smooth and Highlights*. Go to the upper left-hand corner; where it says *Wireframe*, click and select *Smooth and Highlights* from the drop-down menu.

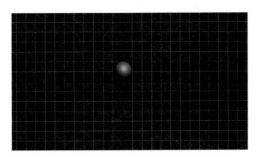

2. At the bottom of the screen on the timeline, next to the key button, click *Auto Key* to put 3ds in automatic keyframe mode.

3. Moving over slightly to the right, open the *Key Filters* menu. Turn off all of them, except for *Position*. This will assure us that while we are animating, we only set keys for position.

4. Go to the *Hierarchy* tab and choose *Pivot*. Under *Adjust Pivot* choose *Affect Pivot Only*. Now use *Select and Move* to move the pivot point to the bottom of the sphere.

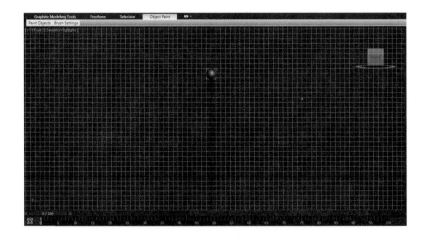

5. At the 0 frame, push the *Set Keys* (big key) button to make our first keyframe. Now, go to the tenth frame and use the *Select and Move* tool to bring the ball to the grid line. Your second keyframe should appear automatically.

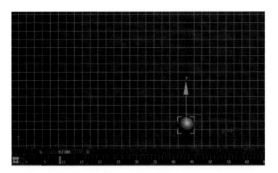

6. Using *Select and Move,* let's set keyframes for all of our bounces.

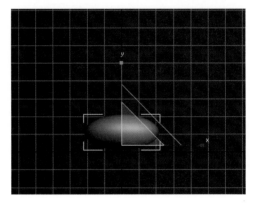

7. Go to the *Key Filters* menu and filter it so that you are only setting keys for *Scale.* Set a key at frame 9. On frame 10, use the *Select and Scale* tool and stretch your ball as shown in the above figure. At frame 11 return it to its original size. Repeat this for each time the ball makes contact with the "ground."

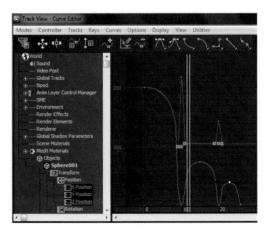

8. With your object highlighted, go to the top of the screen and choose *Graph Editors>Track View – Curve Editor.* The curve editor brings up a similar graph-based editor for adjusting values between keyframes. Just like we did in C4D, adjust the curves as shown in the above figure.

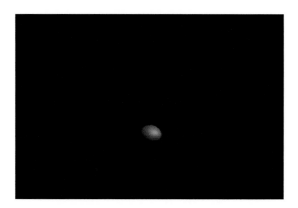

9. To see your animation in action, switch to the *Perspective Viewport* and go to *Views>Grab Viewport>Create Animated Sequence*. This will open the *Make Preview* dialog window.

Animation of a 3D Logo in Cinema 4D

In the following tutorial, we'll add some animation to our 3D logo from Chapter 7.

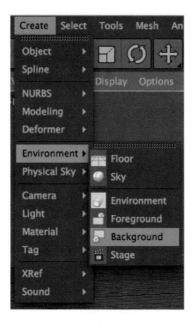

1. Let's begin by setting up a background image for our scene. Go to *Create>Environment>Background* and choose *Background.*

This will create a *Background* object. I created an image in Photoshop for this purpose.

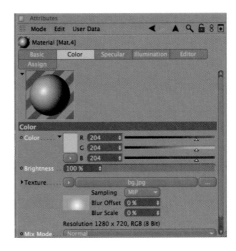

2. Create a new material, and under the *Color* heading import the Photoshop background. Apply the material to the *Background* object.

3. Switch to the *Animation* layout. Turn on *Auto keyframe*. I am going to share with you a secret of the industry, one that works amazingly well and is super-obvious once you've heard it, but

is not necessarily something you'll figure out on your own (at least I didn't). To create a build-style animation, go to the end of the piece and work backward. So at the 60 frame mark (I always leave about 30 frames at the end for readability), we are going to set keyframes for each layer of the logo. So we now know that no matter what we do, it will land where we want it to.

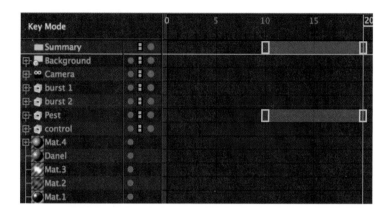

4. At the 10 frame mark, I moved Pest out of frame with the *Move* tool. You'll see that you now have a *Motion Path* and a keyframe at the 10 mark. Then I dragged the second keyframe from the 60 frame mark to the 20 frame mark.

5. I'll do something similar with Control, but this time I make my move at the 20 frame mark, and move the second keyframe to the 30 frame mark. So right after Pest flies in, Control flies in.

6. From the 40 to 50 frame mark, I have burst 1 come in, in much the same way that I did the type. However this time I also add a rotate on the burst at the 40 mark. When it arrives at 50, it will be rotating and positioning.

7. For the second burst, I'll do the same thing as I did with the first, however this time it is from frame 50 to 60. You may want to add in a camera and move with it to finalize your 3D logo.

Character Animation Workaround with the After Effects Puppet Tool

Doing full Pixar-style character animation is a bit beyond the scope of what we are covering with this book, but I do see the value of being able to create some basic character animation for motion graphic artists. Thankfully After Effects has a couple tools, such as the *Puppet* tool and Photoshop 3D, that make some simple character animation a real possibility.

To do true full 3D character animation you would have to get into some pretty complex areas of 3D animation. Basically a 3D character becomes highly unwieldy and unnatural-looking using a parenting-style structure (where one object's movement is controlled by another) or what is more accurately called *forward kinematics* (FK). The preferred, more natural method is referred to as *inverse kinematics* (IK), which means that subordinate child objects will also influence parent objects. To simplify, if someone shakes your hand really hard, your arms move too, so the hand influences the arm. In FK, that doesn't happen, but in IK it does.

To shorten this explanation a little bit, most animators prefer a secondary structure, added to a character model to control it. The process of creating and attaching a control structure to a 3D character is called *rigging*. The term rigging comes from the world of puppeteering. As I said before, this would normally fall outside the scope of this book, but, thanks to After Effects' *Puppet* tool, we can create control points without full-on rigging. Also, thanks to Photoshop 3D we can do this with a 3D character.

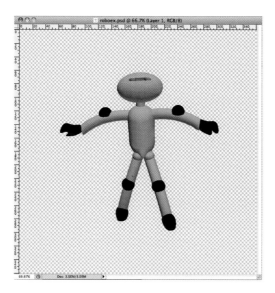

1. For this tutorial, I used the robot character from the cover of this book. I imported the .obj file I made in C4D, and saved it as a .psd.

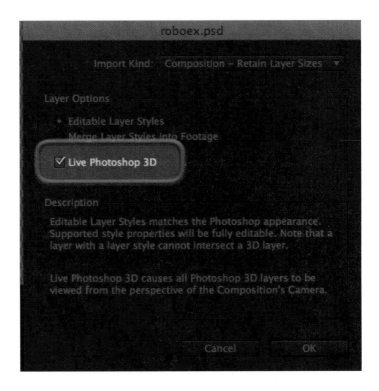

2. Import our roboex.psd into After Effects, and make sure you turn on *Live Photoshop 3D.*

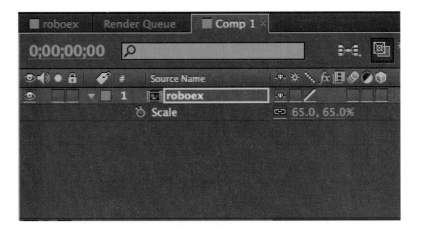

3. Place the roboex composition inside of a new one. We have to nest the composition in order to use the *Puppet* tool on it. Use the *Scale* tool to fit it nicely inside a broadcast-sized composition (for this example I am using the preset *HDV/HDTV 720*).

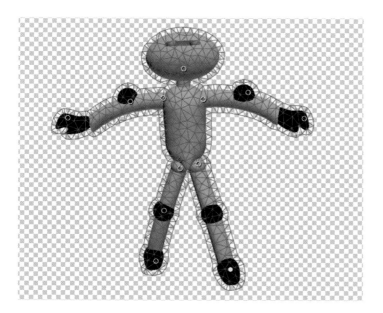

4. Using the *Puppet Pin Tool*, I create points everywhere on the robot from where I think I'll want to make him move.

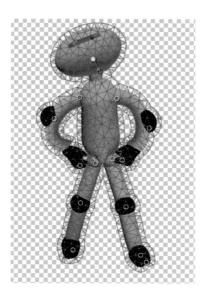

5. Now we can go a little farther down the timeline and experiment with different poses. Here, my guy is getting a little impatient.

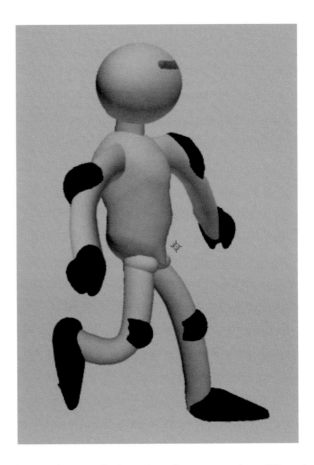

6. This lesson is not designed to be a complete 3D animation solution, but a way to get some simple movement added to a 3D character. Try deleting the puppet effect and changing your character's position and rotation inside the nested comp. I wouldn't recommend changing it during the character animation because it completely skews the puppet effect.

Interview: Dax Norman

Figure 8.2 Dax Norman, painter and 3D animator.

Always looking for new ways of creative expression, Dax Norman is a prolific painter and hopes to define what a twenty-first-century artist can be.

Dax's work in painting seeks to depict multiple realities, coexisting on one plane. "I like to break down images," says Dax, "using their shapes to create other imagery, aided by my subconscious. Because of this process, there appear to be worlds within worlds, yet all are intertwined. The goal is to create a new experience for the viewer upon each successive look." Using this method, multiple examples of personal iconography reoccur in Dax's paintings.

Dax is an avid painter and animator. "My goal is to always enjoy making many kinds of art, so that they all inform and feed off of each other."

In 2008, Dax graduated from Ringling College of Art and Design with a degree in Computer Animation, and graduated from the University of Texas at Austin in 2002, with a degree in advertising. Dax currently resides in Austin, TX, and teaches at the Art Institute of Austin.

What compelled you to start a career in 3D animation?

I have been interested in art all my life, and I wanted to go a very creative route for a career, and I went on a circuitous path to get there. I always loved to draw and paint, and come up with ideas. All of this art that I have done, I had pursued on my own aside from high school art classes, but when it comes to making it a career, I didn't think that I could teach that to myself, so I went to school for it. Once I did, it was like a snowball effect, of loving to do it, and applying the original ideas I had to it. I think it was just how new and interesting the unlimited possibility of what you can create when you have the ideas and the know-how, and just a little bit of technical skill.

Figure 8.3

I find for me that the idea of "unlimited possibilities" is a very driving factor. Where do you see those unlimited possibilities right now?

Making fully immersive worlds, taking one artistic vision from a painting to animation where in spite of the technical differences in production methods, and materials, I can take an idea from one medium to another with a cohesive vision. So I see unlimited possibilities when the technology is no longer an inhibitor to my creativity.

When you first started doing 3D animation, what was the first software package you used and why?

The first animation software I used, that was like a gateway to everything else, was not a 3D one, was Flash. It piqued my interest enough to where I was like "this is fun!" and it got me into a technical skillset that had not developed before. The first 3D software I used was Maya. Basically, because of the school I was going to at the time.

Continued

What software do you use now?

A combination; I still use Maya, and I use ZBrush. I really love ZBrush. For someone whose background is art, and interested in 3D, I highly recommend ZBrush, because it has the most intuitive interface; you can see a direct improvement in what you are doing as you work. I also use Blender.

Figure 8.4

Can you tell me what your workflow is on a project?

I start sketching or with some kind of traditional media. A drawing or painting, something where I have an idea on paper somewhere, where I can look at it. I usually incubate the idea for a really long time. I'll just think about it, what it could be, and that idea will gain momentum in my head about what could be done and where it can go. So by the time I take it into 3D I have a really clear direction. As I think about this idea, I'll doodle little drawings here and there, and build it more in sketches before I take it into 3D.

When you go to the 3D side, what do you do first?

I'll bring in an image plane of one of these drawings, to give me a little bit of a foundation, for what I would want this character or environment to look like. So I can build from a starting point, which often changes, but gives me a nice place to start from.

Then you'll model from the image place, so where do you start your models?

Maya, then I'll create the UV's and develop the whole character before I animate it. I like to try to get the texturing done before I animate it. Then I'll rig and start animating, but I am always tweaking along the way.

Figure 8.5

What's your favorite movie?

My all-time favorite animated movie would be *The Triplets of Belleville*. My favorite live action film would be *The Big Lebowski*.

Can you share a moment when you found something that inspired you and opened up a new level of creativity?

I think it was when I visited Europe for the first time and I saw the work of the world-renowned artists I knew from art history, especially Vincent Van Gogh, and a lot of what I saw in Paris, and the Van Gogh museum in Amsterdam. When I saw these works it was really inspiring to see them in person. I remember when I came back that was a turning point to how serious and dedicated I have been to my artistic craft ever since.

Then later, in 3D software, I saw that you have the power to do things yourself and you don't need a team of people to create an idea. Once you start learning and seeing directly that you can make things, that was really inspiring too.

Also when I first saw the artist Basil Wolverton; he was a comic artist from the 1940s. The thing about his work that was really inspiring, he could have just one drawing that you could laugh at, and almost all of his drawings were funny, and they would include so much texture. It's been my goal to make a model that is appealing that you could look at it, and enjoy it, without it being animated or anything.

What is one skill every 3D artist needs to understand and master as soon as possible?

Once you get to that point where you realize that things don't have to look like they were made on a computer. I think it's really important to me to take something and add a layer of grittiness or a texture, that you can take something and make it look painterly, on a computer or 3D model. Adding multiple texture maps, not just a color map on something, going a little bit further than the basics.

What is your dream project?

My dream project is to complete my own CG animated feature film. I've written it; it's called *Symphony in Sweatch*. Another dream project would be to somehow collaborate with the Coen Brothers. It would be my dream to create an animated sequence for one of their movies.

What have you done so far that you are most proud of?

There is a great feeling just finishing any project. So, what I am most proud of is being able to take so many ideas, and just from a thought, to something tangible that people can watch.

What skills do feel like you are still learning or improving upon?

I have been over the last couple months improving my rigging skills, revisiting the rigging process and not so much relying upon auto-rigs. Creating my own rigs has been freeing since most of my characters aren't the typical humanoid characters. I'm no longer avoiding that part of the process. I figure out how I want to move my characters, and I go in and make that. It has been a weakness for me before, but it's not anymore.

3D TYPOGRAPHY

This chapter will address the main reason why many motion graphics artists will be interested in this book—3D type. After Effects' type animation engine is among the most sophisticated tools there are for type and type animation. In this chapter we'll explore its recently added 2.5D capabilities, but we'll also get into creating true 3D type using Photoshop's Repousse extrusion tool. We'll also look at 3D type with the Zaxwerks 3D Invigorator plug-in and Cinema 4D. So first let's look at the new features of 2.5D type animation in After Effects.

When you create a type layer in After Effects, you have a submenu for text animation options. When you click the *Animate* button, it will open your options for adding animators you type.

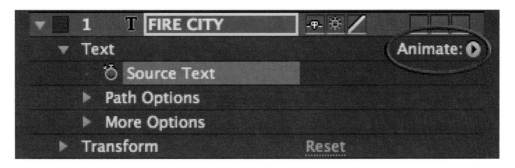

Figure 9.1 The text Animate button.

In the After Effects text tool, an animator is added for any text-specific animation. Since version CS3 there's been a new option under this menu, *Enable Per-Character 3D*. Enabling per-character 3D allows you to have the full 2.5 control of each letter, whereas if you just turn on the 3D switch for your text layer it will put the whole word in 2.5D space.

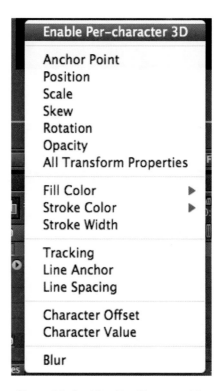

Figure 9.2 Enabling Per-Character 3D.

Using After Effects' Per-Character 3D Text Animation Tools

1. Let's say we are doing a title card for a movie called *Fire City*. Create your title, using a large bold grey typeface.

2. For a subtle touch of depth I use the *Bevel Alpha* effect on the type layer. It can be found at *Effect>Perspective>Bevel Alpha.*

3. Next, I *Enable Per-Character 3D* from the *Text>Animate* menu. I switch to the *Rotation* tool and set the X rotation to about 20

and Y to about 30, so that the type lines up nicely with the line in the background image.

4. Let's add *Text>Animate>Rotation*. Under Animator 1 open *Range Selector 1* and set *X Rotation* to 130 degrees.

5. Open *Range Selector 1*, and over the course of 1 second, animate *Start* going from 0 to 100%.

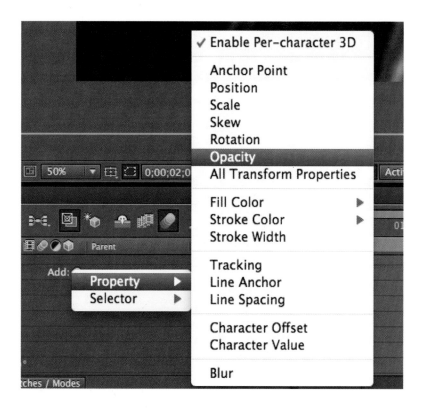

6. Under *Range Selector 1* go to the *Add* button, choose *Property*, and then *Opacity*. Now, set the *Opacity* to 0. You'll see that your type will now fade as the letters rise.

7. To complete this animation, I added a camera to fly across the letters as they rise.

Creating and Animating Volumetric 3D Text with Photoshop's Repousse

Photoshop CS5 introduced the extrusion tool Repousse that we discussed in Chapter 4. Now we will apply it to After Effects.

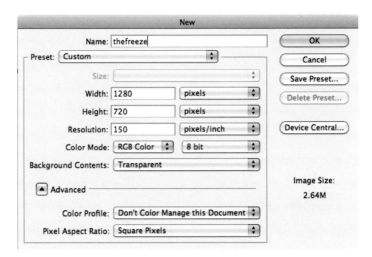

1. Open a new document in Photoshop. Because we are using Repousse, I am setting my ppi higher than I normally would for a TV project. So I change it to 150.

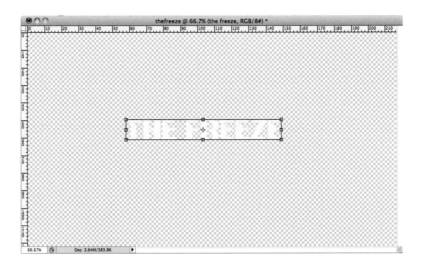

2. Create your type layer. For right now I am going to use white as the type color because when I use 3D materials I'll change it.

3. Apply *3D>Repousse>Text Layer*.

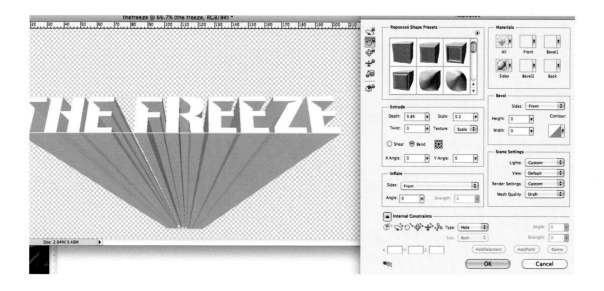

4. First, I'll set the *Depth* of my extrusion really high, over 9. Then I drop the *Scale* to 0.02. This should give a very strong perspective. Then I'll use the 3D *Slide* and *Rotate* to position it like shown earlier. Next I'll click on OK.

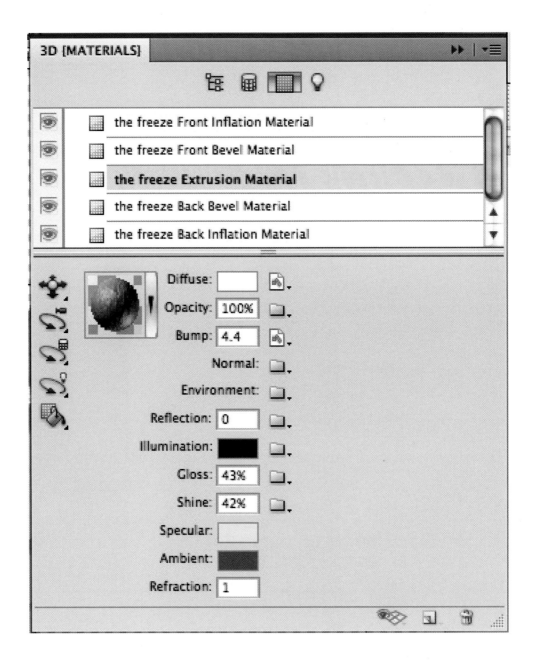

5. Now let's use some custom materials to get the look we want. Go to the *3D Materials* panel. Highlight "the freeze Extrusion Material." Under *Diffuse* load the provided icetexture.psd. Also load that material for *Bump*. Set the *Bump* to about 4.5. Increase both *Gloss* and *Shine* over 40.

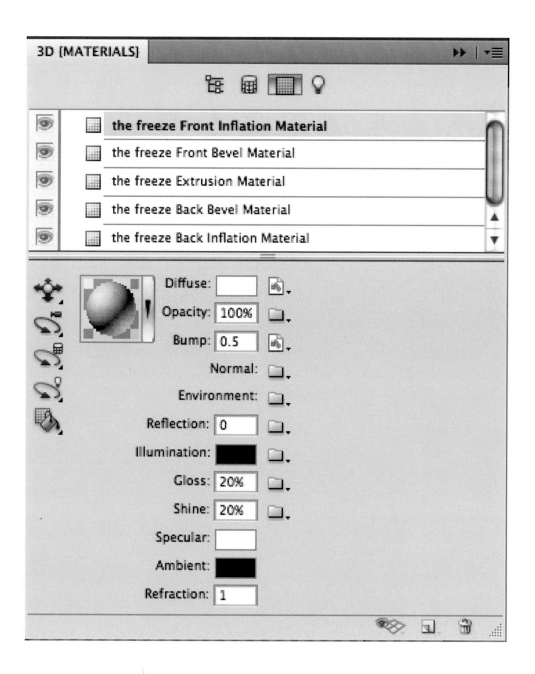

6. Now let's highlight the "the freeze Front Inflation Material."
Under *Diffuse* load the provided icetexture2.psd. Also load that
material for *Bump*. Set the *Bump* to about 0.5. Make both *Gloss*
and *Shine* about 20.

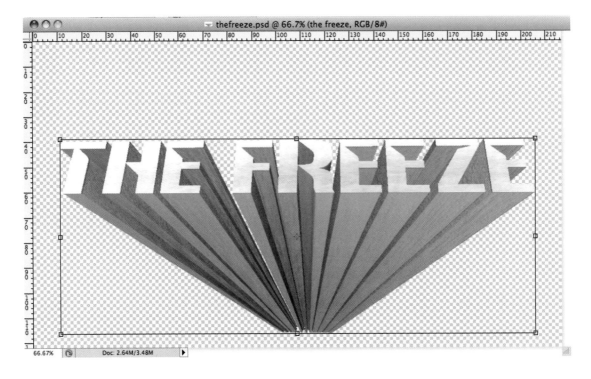

7. Save our file for After Effects.

8. Import our file into After Effects as a *Composition*, and make sure that *Live Photoshop 3D* is on. You should notice a big problem. It looks awful. Now, flipping quality switches in AE will do no good; we actually have to go back to Photoshop to fix this. Highlight the bottom "freeze" content layer and press Ctrl/Cmd-E to open the file in Photoshop.

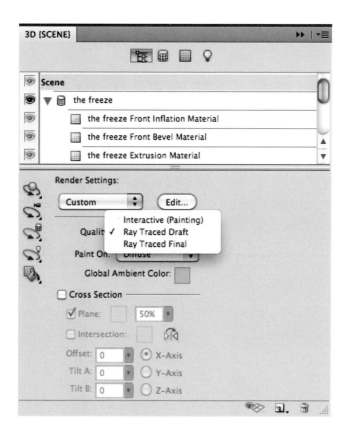

9. In the 3D Materials menu, under *Quality*, switch it to *Ray Traced Draft*. Give it a second or two, and you'll see it render a higher quality version of your image. Save your file.

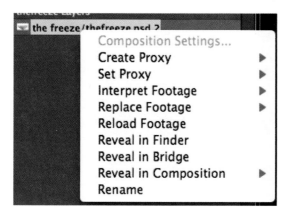

10. Now return to After Effects. It should update, but if not, right-click the Photoshop file in the Project panel and choose

Reload Footage. It looks much better now, but After Effects may start to move like a glacier depending on your RAM. You can also just wait until you've completed all your keyframing before you execute this step.

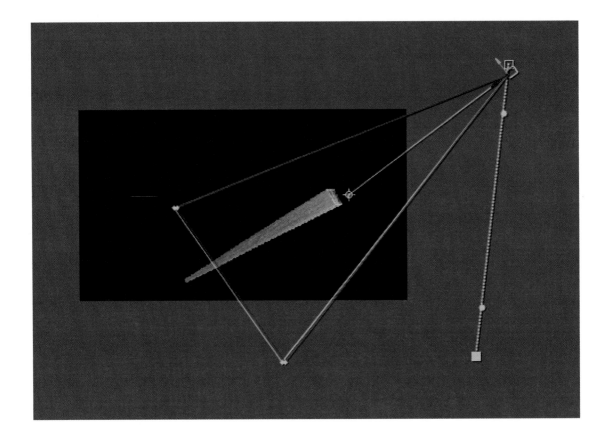

11. I switch to the *Left* camera view, and animate the *Camera* layer's position to move from below the type to above.

Creating 3D Type in Cinema 4D

Repousse in Photoshop and After Effects is a bit slow and limited. For many processes it makes more sense to use a full 3D host like Cinema 4D (C4D). The first tutorial is on 3D type in C4D. We'll look at creating text in 3D and then we'll animate it.

1. There's actually a couple of ways to make text in C4D, but the easiest method is to use the *MoText Object* from the *MoGraph* menu.

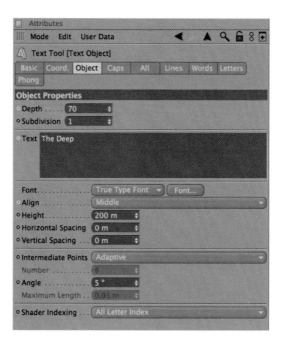

2. Under the *Attribute Manager* for the *Text Object*, go to the *Object* tab and increase the *Depth* to 70. Enter the text you

want it to display and set the font by clicking the *Font...* button.

3. Once you have set your type, you can go ahead and add a material to your text. I also went ahead and made a ground plane. You may also go ahead and animate your word, however you will be limited to using the word as a unit, and will not be able to do anything with the separate characters.

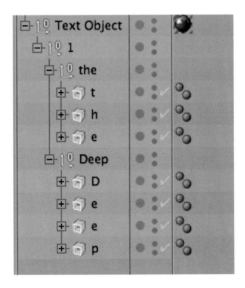

4. To animate the separate characters, you'd need to press C on the keyboard and convert the text to an editable object.

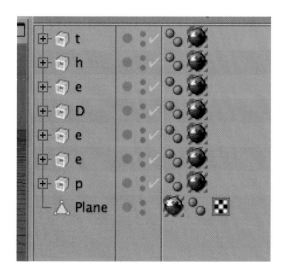

5. I expand my different object groups in order to make them a bit more manageable. Once you have expanded an object group, you delete the old one, since it no longer contains anything like an empty folder.

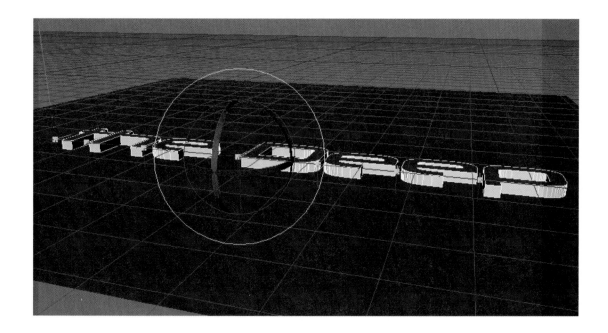

6. Put your timeline in *Automatic Keyframe Mode*. Select all your objects and set a keyframe at the 60 frame mark. At 0,

use *Rotate* and set them at –90 degrees (or you can open the *Coordinate Manager* and on the *P* axis type –90 and then click *Apply*). If you scrub through the timeline the whole word should stand up.

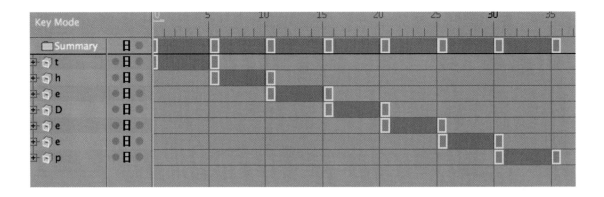

7. To animate our letters standing up, I drag the end keyframe for the T to the 5 frame mark. And then I do the same for each letter one at a time rising, so h is from 5 to 10 and so on.

8. I added a target light to complete the look of the scene. Feel free to try adding some motion blur to give the scene a professional, polished edge.

Exploding Type in Cinema 4D

This is quite popular because Cinema 4D's take on this is easier than I've seen in other software. The following tutorial shows you how you can blow up some type.

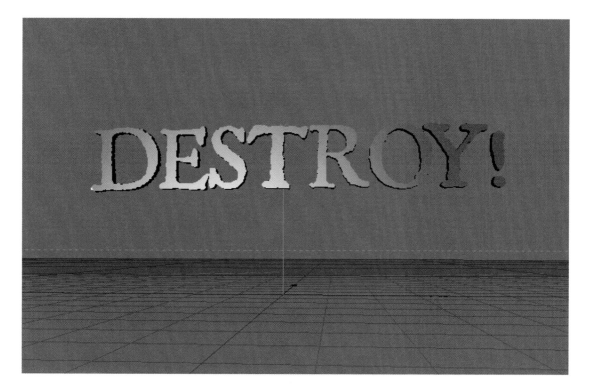

1. Set your type.

2. From the *Deformation Objects* menu select *Explosion FX*. Place it inside of the text object.

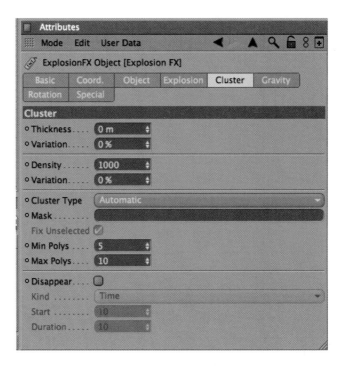

3. In the *Attribute Manager* for the *Explosion FX* object, open the tab for *Cluster*. Set the *Thickness* to 0.

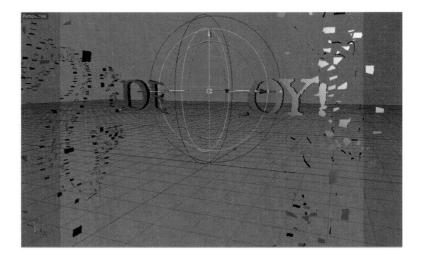

4. Move the *Explosion FX* object into the middle of the word; you'll see that it is already doing its thing, so you may want to start with it in a much calmer place. In the *Attribute Manager* under *Object*, scrub down the value of *Time* to 0%.

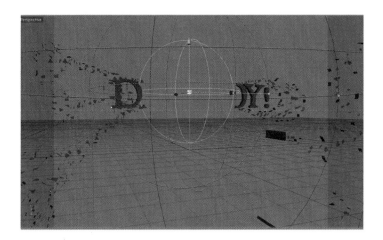

5. Now put C4D into *Automatic Keyframe Mode* and create a keyframe at the 0 frame. Go the 60 frame mark. Increase the *Object>Time* in the *Attribute Manager* (it's also the green ring of the *Explosion FX* controls). Also increase the *Explosion>Blast Range* (red ring) and *Gravity>Range* (blue ring).

6. There is a lot to experiment with using this very cool effect. Try adjusting the various options to find a look you like. Also, the

Explosion FX controls themselves can be animated throughout the word, so that's another variation in this effect that works great.

Creating 3D Type in 3ds Max

1. To create a text object in 3ds Max, start by going to *Create>Shapes>Text*. Choose your font and a size, and set leading and kerning in the *Parameters* menu below. Type what you would like your text to read.

2. Click the viewport to place your type.

3. Go to the *Modifier* stack and apply *Extrude*.

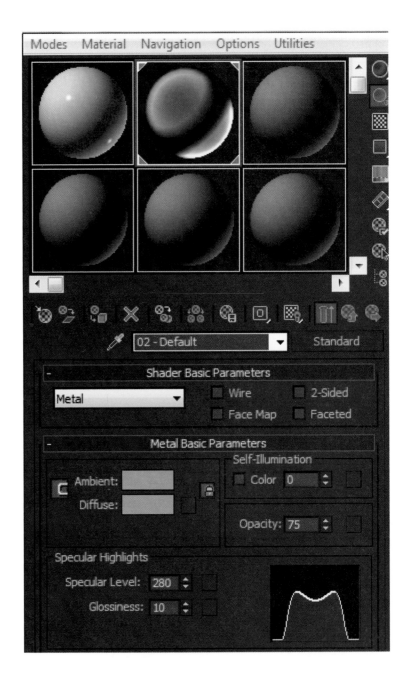

4. Now that our type is 3D, let's add a better-looking material to the object. Open the *Material Editor,* and choose *Metal* from the drop-down of *Shader Basic Parameters.* I set the *Ambient* and *Diffuse* colors both to a light blue. Set the *Specular Level* to 280 and the *Glossiness* to 10. Set the *Opacity* to 75. Now apply the material to our type.

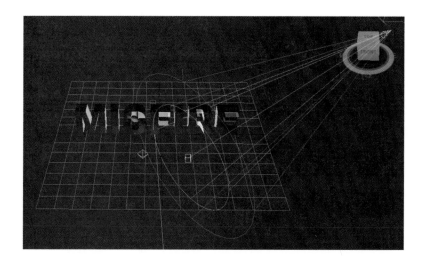

5. Add a *Target Spotlight* from the *Lights* menu. I placed mine behind the type to give it a nice backlit effect. Animate it traveling through the letters. Also, add an *Omni* light directly below the type. Adjust the distance. We are trying to light the front of the text, without losing the highlights from the light behind.

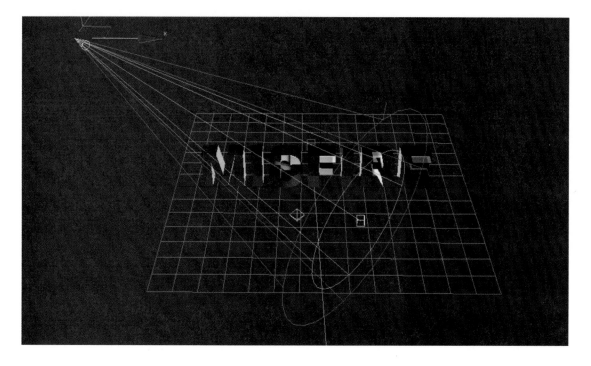

6. Add a camera and have it zoom back from the type.

7. When finished it should have a very cool film-noir feel to it.

Text Animation with Zaxwerks Pro Animator

For a long time the quickest way to get usable 3D objects into After Effects was to use the Zaxwerks3D Invigorator plug-in. Even though times have changed, there are still some very useful things that can be done with Zaxwerks' Pro Animator, 3D Invigorator's big brother. Zaxwerks was always quite different from most other AE plug-ins in that it was more like a complete 3D modeling package that just lived inside of an AE plug-in. When you were done with what you wanted it to do, you had an After Effects layer. So, it's no big surprise that Pro Animator is also available as a stand-alone product now.

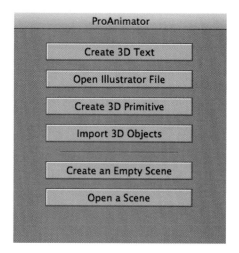

1. From the main menu, choose *Create 3D Text*.

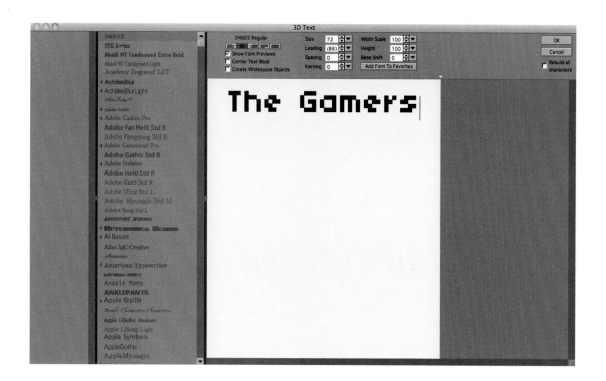

2. In the 3D Text menu type out your text, and edit it the way you
   like; when you are done press OK.

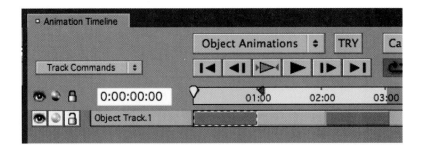

3. Zaxwerks has something of an odd take on keyframes. Rather than using the normal lovable little diamonds and squares we are used to, Zaxwerks opts for these bars, which will take a little getting used to. Go to the two-second mark and double-click. It will create another dark blue bar like the one we already have. There's a light blue bar in between. That light blue bar is where our animation will occur. Highlight the first dark blue area before moving to the next step.

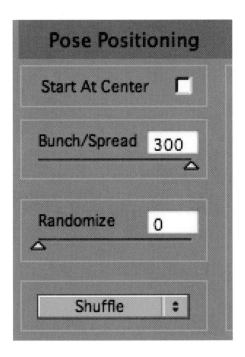

4. Under *Bunch/Spread* increase it to 300. You'll see it do something like a tracking animation.

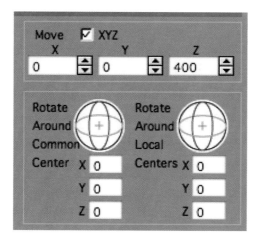

5. Over in the *Move* section, check on *XYZ*. Under *Z* make the value 400.

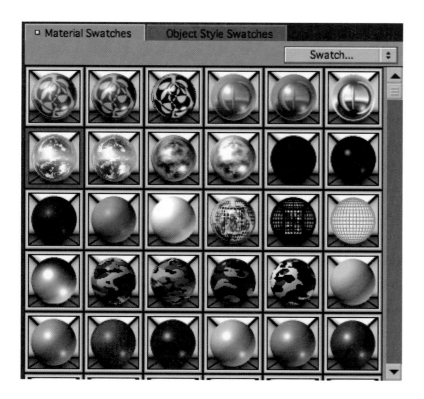

6. One thing that the Zaxwerks product line is well known for is plenty of preset materials to choose from. Go to the *Material* view. I used the preset for *Smooth Ice* on my type.

7. Preview your animation. Zaxwerks' Pro Animator is a powerful set of tools for those of us who need a few quick 3D touches.

CINEMA 4D MOGRAPH TOOLS

Cinema 4D (C4D) is a big piece of software, with tons of powerful tools. Packaged with C4D is a pretty important module called MoGraph, designed specifically for the needs of the motion graphics designer. MoGraph 2 arrived with Cinema 4D R11.5, and expanded the original toolset by adding *PolyFX, MoSpline*, and *MoDynamics*. MoGraph adds a series of procedural modeling and animation tools designed to collect many of the things that motion graphics designers need in one place. The best way to see the MoGraph toolset is to switch to the *Broadcast* layout.

Figure 10.1 The MoGraph toolbar.

Cloner Tools

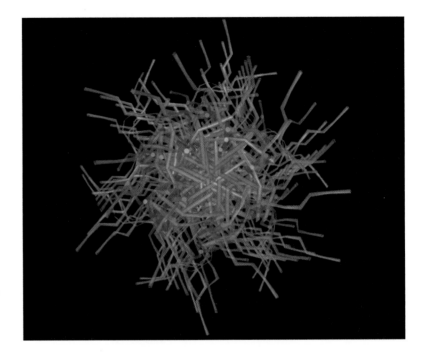

Figure 10.2 C4D's *Cloner* tools make complex abstractions a fairly quick and painless process.

The first MoGraph tool we'll be looking at is the *Cloner Tool* and *Cloner Objects*. A *Cloner Object* is an object where you have specific controls over the number of duplicates of a single object and how it is patterned. Anytime you use one of the *Cloner Tools*, whichever object you click becomes part of a *Cloner Object*, including other *Cloner Objects*.

Figure 10.3 The *MoGraph Selection* and *Cloner* tools.

The first tool on the MoGraph toolbar is the *MoGraph Selection* tool, which allows you to select *Cloner Objects, Matrix Objects, Fracture Objects*, and *Text Objects*. Even though you can already do this with the *Live Selection Tool*, the *MoGraph Selection* can create selections to which you can restrict effectors. The next three tools are the *Cloner Tools*. They work like a select and copy tool. First the *Linear Cloner* tool creates duplicates in a straight line. The *Radial Cloner* creates circular patterns and the *Grid Cloner* tool creates, you guessed it, a grid.

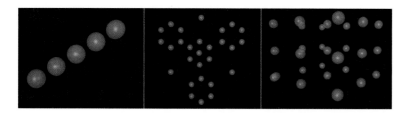

Figure 10.4 *Linear, Radial,* and *Grid Cloner Tools* in use.

The next group of tools on the MoGraph toolbar is the MoGraph objects set. Each of these tools are designed to create an object that corresponds to some MoGraph function.

First is the *Cloner Object. Cloner Objects* are created when you use any of the *Cloner Tools,* but if you'd prefer to use their functions without the quick complexity of the *Cloner Tools,* you can just make a *Cloner Object* and adjust its parameters individually. The next tool is the *Matrix Object.* A *Matrix Object* is just like a *Cloner Object* except that it doesn't create anything that can be rendered. It basically applies an orthogonal matrix to the object rather than duplicating it, making it better for processing deformations than a *Cloner Object. Matrix Objects* can also be used in conjunction with *Thinking Particles.*

Cloner Object Tutorial: Modeling an Abstract Sun

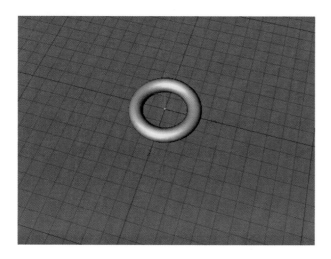

1. Create a *Torus* with a *Radius* of about 200 m. Make a new *Material* (that's yellow), and be sure to turn on *Glow.* Apply to the *Torus.*

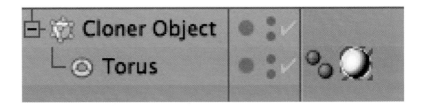

2. Create a *Cloner Object.* Put your *Torus* in the *Cloner Object.*

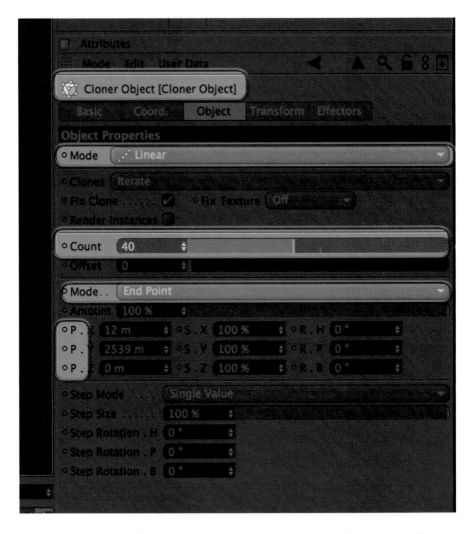

3. In the *Cloner Object Attribute Manager,* set the *Mode* to *Linear.* Make the *Count* 40. Change the *Mode* below *Offset* to *End Point.* Under there adjust the *P*'s Y axis so you have a nice tower of *Tori.*

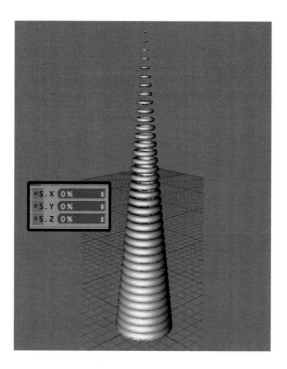

4. Still inside of the *Cloner Object Attribute Manager,* move over to the *Scale* percentages and make them all 0. Because we are using the *End Point* mode, over the course of our duplication, they gradually decrease in size.

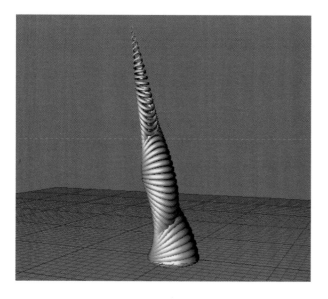

5. Adjust the *Position* and *Rotation* values until you have something like the above figure.

6. Using the *Radial Cloner Tool,* set the *Count* to 15 in the *Attribute Manager.* Click your *Cloner Object* and drag to make 15 duplicates of your original *Cloner Object.*

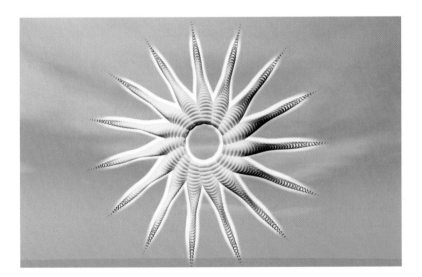

7. To complete my scene I create a *Sky* from the scene menu, and preview.

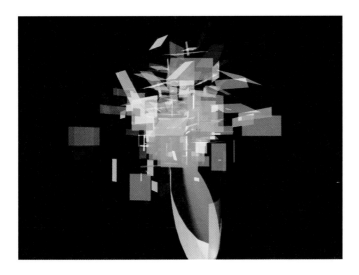

Figure 10.5 A capsule object within a Fracture Object with the Random Effector applied.

Fracture Objects

Returning to our MoGraph toolbar, the next object to the right is the *Fracture Object*. A *Fracture Object* is not just another exploder type tool. Rather, *Fracture Objects* allow you to apply *Effectors* (we'll get into a deeper discussion on *Effectors* later in this chapter) to a desegmented object and give you a significant level of control over your deconstructed 3D objects. In Figure 10.5 I've taken a capsule and applied the *Fracture Object* and *Random Effector* to it. In the following tutorial, I am placing a *Text Object* inside a *Fracture Object* to do another exploding type move.

Using a Text Object with a Fracture Object and Random Effector

1. Under the *Spline* objects, choose *Text*. Create your type, and then apply *Extrude NURBS*. This is the method that folks

used to create type before MoGraph. For this lesson using this method is a little faster.

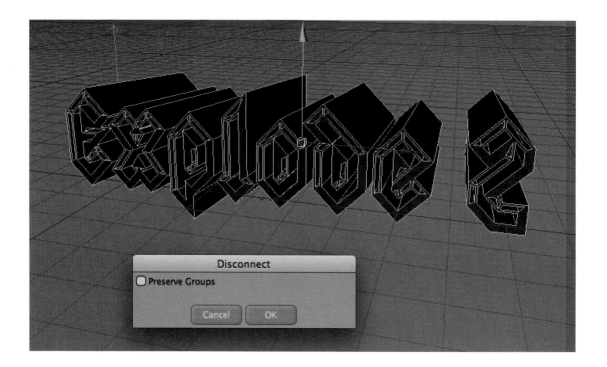

2. Press C on the keyboard to switch to an editable object. Select the *Polygon Tool* and press Ctrl-Cmd-A to select all polygons. Now, go to the *Functions* menu and choose *Disconnect*. When the *Disconnect* dialog opens, turn off *Preserve Groups*.

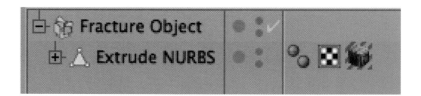

3. Create a *Fracture Object* from the *MoGraph* toolbar. Make the *Extrude NURBS* object a child of the *Fracture Object*. With the *Fracture Object* settings active in the Attributes Manager, go to *Object* and change the *Mode* to *Explode Segments*.

4. At this point it will look as if not too much is happening. The thing that makes *Fracture Object* controllable is the interaction it has with *Effectors.* On the *MoGraph* toolbar choose a *Random Effector.*

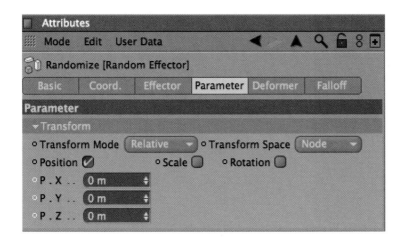

5. Place the playhead at the 60 frame mark. Turn on *Automatic Keyframing.* Make a keyframe. In the *Random Effector* go to *Parameter* and set all the *Positions* to 0.

6. Set the playhead at the 0 frame. Make the X and Y *Positions* 1500 m. Scrub through your timeline and you'll see your letters forming into words.

7. To make a complete scene create a *Background Object*. Create a new *Material* and click *Load Material Preset*. MoGraph even has its own preset materials. I used *Animated Backgrounds>Electric Clouds.*

Instance Objects

The *Instance Object* is the next tool in our collection of MoGraph tools. *Instance Objects* behave in a similar way to the *Echo* effect from After Effects. As you animate an object between two keyframes, it will leave a trail of copies in its wake. They will disappear over time. This is great for creating very interesting chain patterns, quickly and easily. It will also be great for working with type.

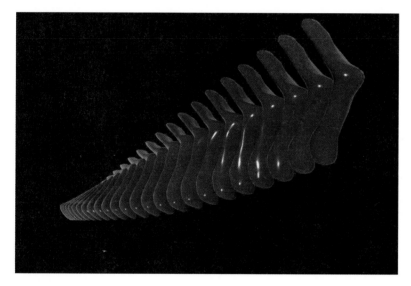

Figure 10.6 The *Instance Object.*

Using Text with an Instance Object

1. I created a *Text* spline that I extruded with *Extrude NURBS*. Switch the text to editable mode by pressing C.

2. Create an *Instance Object.* Make the *Extrude NURBS* a child of the *Instance Object.*

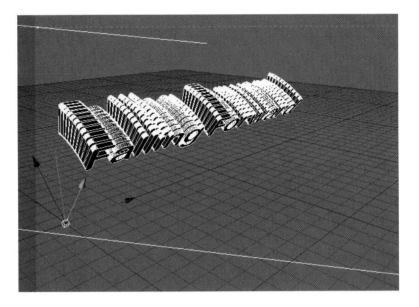

3. With your *Instance Object* selected, put C4D into *Automatic Keyframe* mode. Set a keyframe at 0 for *Rotation.* Go to 50, and rotate the type moving forward. To give it that nice falling forward motion, I moved the pivot point well below the type.

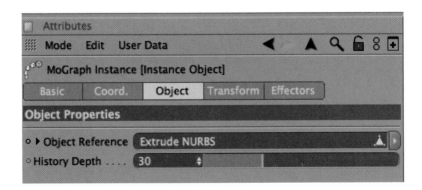

4. Turn off *Automatic Keyframe Mode.* Open the *Attribute Manager* for the *Instance Object.* Raise the *History Depth* to about 30 to make more instances of our text to follow the initial one.

5. I think I have spent entire days playing with *Instance Objects*. This effect we executed here will work both as a type animate on and off effect. I added a light, a ground plane, and background object to complete this effect.

Text Objects

As we have discussed earlier, there are two methods for creating text in C4D. You can use a *Text Spline* and *Extrude* to create type or you can use the MoGraph *Text Object*. A *Text Object* has a number of advantages: (1) it will already be extruded, and (2) you can apply MoGraph *Effectors* to it. This is the closest you will come to the AE Text animation engine that is so popular in the industry in true 3D. The only time a *Text Object* will work against you is if you have an effect that will require you to break it down into its original mesh. It's much more of a hassle to get to it from a *Text Object* than from an extruded *Text Spline*.

Figure 10.7 A *Text Object* with a Random Effector applied to Position, Scale, and Rotation.

Per-Character Type Animation Using Text Objects and the Step Effector

1. Using a *Text Object*, create your type.

2. From the MoGraph toolbar, create a *Step Effector*.

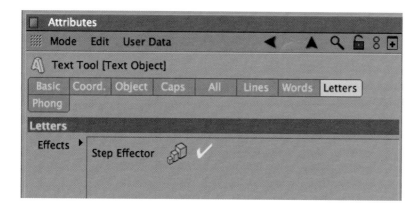

3. The *Step Effector* should appear on the *Effects* list in the *Text Object's Letters* tab. At any point in time, an *Effector* can be dragged to any *Effects list* that appears within a *MoGraph* object.

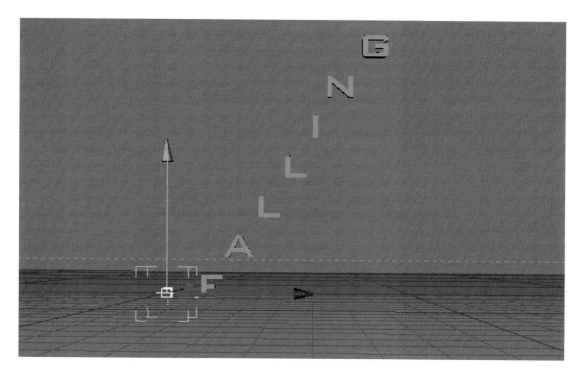

4. Go to the *Step Effector's Parameter* tab. Turn off *Scale*. Turn on *Position*. Increase the *Y Position*. You'll see the letters start to rise, but not consistently. I'll show you why.

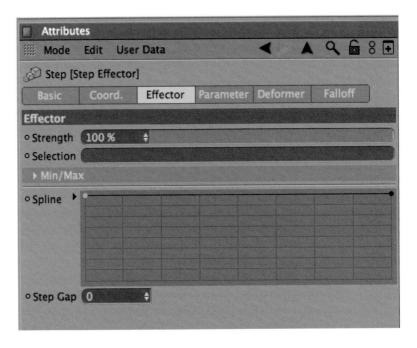

5. Go to the *Effector* tab. Take a look at the *Spline*. See how it is a ramp? Pick up the first point and drag it up so it is even with the second point. Now you'll see a consistent rise in the type.

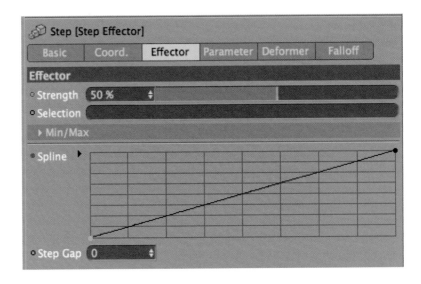

6. Turn on *Automatic Keyframe* mode. At the 25 frame mark, I drop the *Strength* to 50% and the first point of the *Spline*.

7. At the 50 frame mark, drop the *Strength* to 0%. The type will land as you would want it to for the final presentation.

8. To add another level of dazzle to this, I decided to add a *Rotation Parameter*. At the 0 frame, turn on the *Rotation* parameter. Set the degree of the three axes to any number you want (in this case the higher the better). The existing animation of the *Step Effector* will take care of the rest.

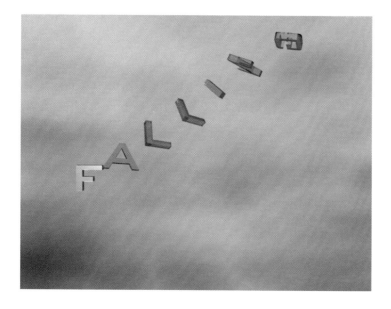

9. Now the letters will be falling into place. I applied a glass style material to the letters and added a slick background layer. This effect looks great and demonstrates the possibilities for animated text with MoGraph in C4D.

Tracer Objects

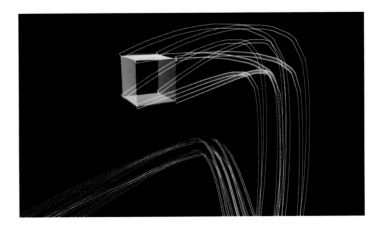

Figure 10.8 The MoGraph *Tracer Object.*

Often what a motion graphics project needs from 3D is that little bit of edge to be slicker than the crowd and stand out. The *Tracer Object* does exactly that. It works in a similar way to the *Instance Object*, but rather than making copies of the object as it goes through the animation, the *Tracer* creates lines that extend from each vertex of the animated object. These lines can be extruded. It's great for creating those wonderful organic vine type effects like we did previously in *Particular*.

Creating a Futuristic Comet Using the Tracer Object

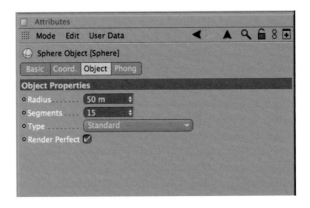

1. Let's begin by creating a *Sphere* object. Since a sphere has tons of points, we need to simplify it a little. Set the *Radius* to 50 m and *Segments* to 15.

2. Create a new *Tracer Object.* In the *Attributes Manager,* under the *Object* tab, it will automatically set the *Trace Link* to *Sphere.*

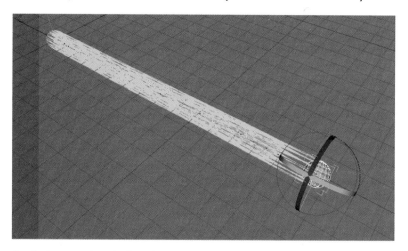

3. Animate the *Sphere's Position* and *Rotation.* You'll see lines emitting from behind the *Sphere.* These come from the *Tracer.* But we don't have anything visual from them yet.

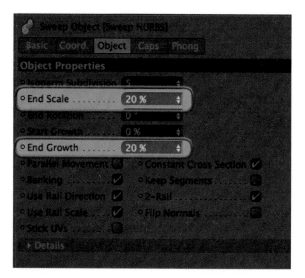

4. Create a *Sweep NURBS* object. Make the *Tracer Object* a child of the *Sweep NURBS* object. In the *Sweep NURBS Attribute Manager,* set the *End Scale* and the *End Growth* to 20%.

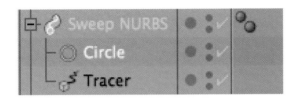

5. A *Sweep NURBS* object needs at least two splines. So, create a *Circle* spline, with a *Radius* set to 10 m. Also make this spline a child of the *Sweep NURBS*.

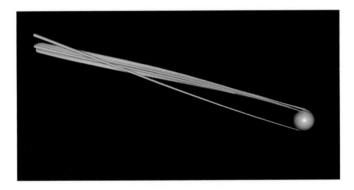

6. Scrub through your timeline and you should see something like the image in the above figure.

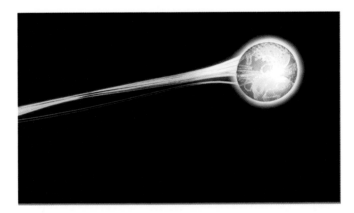

7. To complete my comet, it's now just a matter of materials. I found two perfect materials for this in the presets. For the *Sweep NURBS* I used *Broadcast Edition>Effects>CSTools-Flux-Electric Blue*. For the sphere I used *MoGraph>Glossy Transparent>Frozen*.

Spline Mask Objects

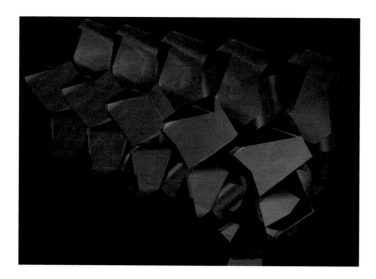

Figure 10.9 *Spline Mask Objects* combined with *Cloner Objects*.

Spline Mask Objects are fairly easy to understand and use. Basically they comprise two splines, combined either by adding or subtracting the overlapping areas, in a similar way to how Adobe Illustrator's Pathfinder options work. Then they can extrude using an *Extrude NURBS Object*.

Using a Spline Mask Object

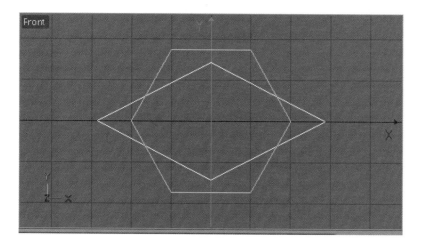

1. In the front viewport create two splines that overlap.

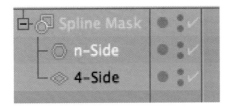

2. Place both splines inside a new *Spline Mask Object.*

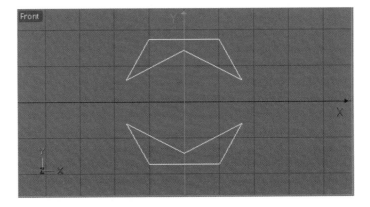

3. In the *Attribute Manager* for the *Spline Mask Object,* switch the *Mode* to *A Subtracts B.* You'll see how the shape changes.

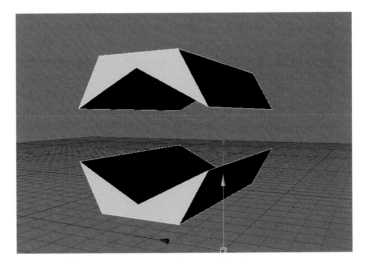

4. Create an *Extrude NURBS Object,* and make the *Spline Mask Object* a child of the *Extrude NURBS Object.*

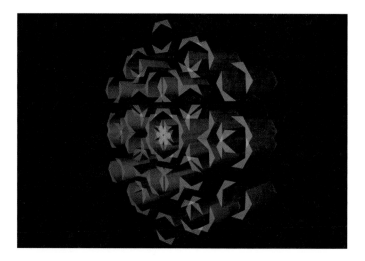

5. I used the *Radial Cloner Tool* on the shape twice to get a fairly complex mechanical-flower-style shape. I also added a material.

6. To animate this object, I added the *Shader Effector*. I connected to the *Effectors* tab of the *Cloner Object*. Then under *Attribute Manager >Parameters*, I enabled *Scale, Position,* and *Rotation.* Experiment here; at this point everything you do will probably look great.

MoSpline Objects

MoSpline Objects arrived with R11.5 and MoGraph 2 and are often considered among the most useful and exciting tools in the MoGraph package. *MoSpline* is possibly the best tool on the

market for creating those great organic line effects. It is so easy to get absolutely gorgeous designs with this tool that it is no wonder this effect has become very commonly used all over television. In the following tutorial I'll take you through a standard setup procedure for creating 3D vines with *MoSpline*.

Creating 3D Organic Vines with MoSpline

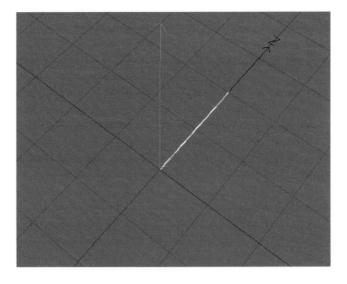

1. Create a new *MoSpline* object.

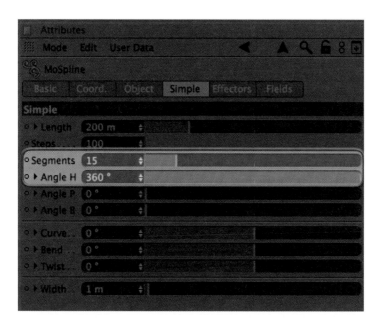

2. In the *MoSpline Attribute Manager,* increase the number of *Segments* to 15. Under *Angle H* increase the degree to spread out the segments.

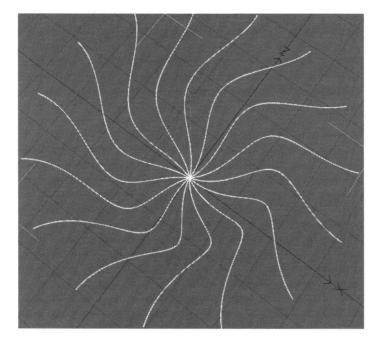

3. Go farther down the *Attribute Manager* for *MoSpline* and adjust the *Curve, Bend*, and *Twist* to make the shapes more interesting.

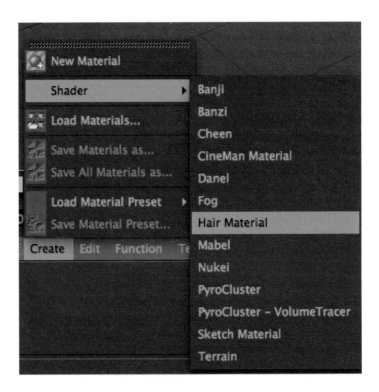

4. Go ahead and preview. Don't see anything? That's right, nothing we have done is visible yet. Go to the *Material Manager* and create a new *Hair Material*. Apply it to our *MoSpline*.

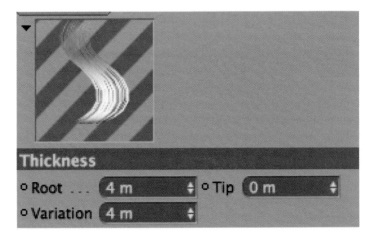

5. In the *Hair Material's Attribute Manager*, go to the *Thickness* tab. Make the *Root* 4 m and the *Tip* 0. I also made the *Variation*

0 as well. I also went to the *Color* tab and added more color stops to the gradient.

6. Experiment with a variety of different controls when animating to see the range and beauty of what a MoSpline can do for your motion graphics projects.

MoGraph Deformers

Figure 10.10 The MoGraph deformers, Extrude and PolyFX.

In addition to our MoGraph objects, MoGraph also provides us with some specialized *Deformers*. Figure 10.10 shows our MoGraph deformers. The MoGraph deformers may recall other C4D tools. However the MoGraph deformers can be used in conjunction with *Effectors*, and as you have seen that is a huge advantage.

Figure 10.11 A plane with a Displace deformer applied to create a water surface.

Extrude Deformers

Extrude deformers are a lot like other extrude methods but they have a few more levels of complexity. First, you can use *Effectors*; second, they can extrude the polygons of primitives, and you can extrude along a spline. They can quickly extrude into excited and complicated shapes. In Figure 10.12, the image I made started with a *Sphere* with the *Extrude Deformer* as a child. I drew a custom spline shape and set that to the *Sweep Spline* in the *Extrude Deformer's Object* tab in the *Attribute Manager*.

Figure 10.12 Object created using an Extrude deformer.

PolyFX Deformers

Figure 10.13 The PolyFX deformer applied to a sphere with the Spline Effector.

The *PolyFX* deformer is the last on our list of deformers that come with the MoGraph bundle. The PolyFX deformer can be used as another exploder-style effect, but it has some differences to other similar functions of C4D. It's a deformation object that will view segments or splines as clones and your applied Effectors will treat them that way. Another great thing about the PolyFX deformer is that the more segments you use on your object, the more fine your object can be exploded, without switching to an editable object.

Figure 10.14 A sphere is blown into confetti using PolyFX, a Spline Effector, and Random Effector.

To make use of the *PolyFX* deformer, create an object and make PolyFX a child. You won't see anything until you add an effector. In Figure 10.15 you'll see a sphere with PolyFX and a Random Effector.

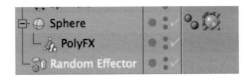

Figure 10.15 You need an effector to get results from PolyFX.

Effectors

Figure 10.16 Effectors (top to bottom): Group, Plain, COFFEE, Delay, Formula, Inheritance, Random, Shader, Sound, Spline, Step, Target, Time, and Volume.

Throughout this chapter, we have been using *Effectors* in conjunction with our MoGraph objects and deformers. Think of them as specific effects designed to easily control aspects of our MoGraph objects, points, clones, and such. Many of the deformation objects and MoGraph objects need an effector to show results. Effectors can be grouped into two major categories: (1) those that define initial values for clones and transform them based on clone parameters, and (2) special effectors, which need something specific (like a spline or animated object) to carry out an action.

The ones that are used to define clone values are:

- Plain: Assigns values in the parameter tab to each clone
- Formula: Assigns values based on a formula
- Random: Assigns random number values
- Shader: Creates values based on a texture's grayscale values
- Sound: Values based on audio frequencies
- Step: Supplies consecutive values
- Time: Values increase over time

The effectors that work with special properties such as a supplied spline or animated object are:

- Group: Combines multiple effectors into a single effector
- Delay: Sets a smooth transition for the beginnings of effector parameters (other effectors required)
- Inheritance: Follows the user-supplied reference object
- Spline: Clones follow the shape of a user-supplied spline
- Target: Orients clones to an object (like a camera)
- Volume: The influence of an effector is limited by the size of a user-supplied object
- COFFEE: The most complex of the effectors in that it uses the C.O.F.F.E.E. language that can be used to write C4D plug-ins; allows users to take full control of the behavior of clones
- Python: Similar to the COFFEE effector, allows users to write Python code to take full control of clones

The MoGraph package is an essential addition to the C4D package for motion graphics designers, and is not limited to just the areas addressed in this chapter.

Interview: Erica Hornung, VFX/Matchmove Artist

Figure 10.17 Erica Hornung, VFX/matchmove artist.

Erica Hornung has worked as a Senior Matchmover and Technical Director at the Academy Award-winning Hollywood visual effects studios Sony Pictures Imageworks and Rhythm and Hues, among others. Her film credits include the *Spider-Man* series, *Lord of the Rings: The Two Towers*, the *Matrix* series, *The Polar Express*, and most recently, *Captain America: First Avenger* and *Hansel and Gretel: Witch Hunters*, along with other blockbuster movies.

Erica received a BS in Architecture from Georgia Tech and an MA in Film from Georgia State University. Both have served her well in her career, exposing her not only to complex 3D visualization, but also film production techniques.

Erica is one of the lucky few who pursued a childhood dream and achieved it. She lives in Los Angeles with her increasingly demanding cat, Harper.

She is the author of *The Art and Technique of Matchmoving*, published by Focal Press in 2010.

What compelled you to start a career in 3D animation or art?

I saw *Star Wars* on December 24, 1977. I knew pretty much immediately that I wanted to do "that."

What was the first 3D software package you used and why? What do you use now?

I used 3DS (Note: 3DS, not Max! Wayback machine, activate!) in grad school for a project. We also got Alias 6 at school, which I played with. I got an internship at Cartoon Network, and we were actually beta testing the first Maya release there. Now, I use Maya almost exclusively, save for studios that use proprietary software.

Describe your workflow on a project.

On live action feature films, I'm usually a matchmover, which is at the beginning of the pipeline. I approach every matchmove shot the same: analyze the camera or actor motion and identify the extremes. I set up the camera and character poses on the first frame, then see how it generally all fits together. Next, if any 2D tracking is needed, I'll do that. Then, it's a matter of bending the CG camera to my will. When animating 3D stand-ins, I do several passes where I only watch the actor's core, then add in more refined movements as I go.

In broadcast design, I usually start with fonts. Fonts influence the look and feel of a piece so much! I also go to sites like motionographer.com, watchthetitles.com, and YouTube to look for examples that might be relevant to the project.

What is your favorite movie?

Star Wars. Though of course *Empire* is the superior film—I love *Star Wars*.

Can you share a moment when you found something that inspired you, and opened the door to a new level of creativity?

I had refused to touch computers as a kid and in undergraduate school. My dad is a programmer, which might have something to do with it. But when I saw *Jurassic Park* and *Terminator 2*, I realized this CG thing might have legs. So I concentrated on teaching myself throughout grad school, and loved it.

Continued

Figure 10.18

Figure 10.19 Stills from *The Art and Technique of Matchmoving.*

What is one skill that you think every 3D artist needs to understand and master as soon as possible?

The curve editor! I'm always so surprised when students and even coworkers not only don't use the graph editor, but have never even seen it!

I'd also like to see people making better use of the help docs and learning to data mine better to solve problems rather than waiting for someone else to solve their problem.

Where do you look for inspiration?

I look everywhere! Books, especially children's books, are great. Old record covers are really inspiring. Current shows and commercials—I watch them over and over! I also have a camera with me everywhere I go, so I can shoot inspiring sights, from flowers to graffiti to architecture. I also get RSS feeds from various design sites. Whenever I see something I like—or really hate—I pull it and put it in an "inspiration" file. They serve as my screen saver and desktop images.

What is your dream project?

That's a good question. I've worked on so many great films and promos, it's almost like I've already done a dream project. I think a dream project for me would be VFX sup for an independent film. It would be cool to have creative control over my work. Likewise, art director at a motion design shop would be pretty dreamy!

What project are you most proud of?

I was pretty proud of my thesis project, not because it was so awesome (it wasn't), but because it was so insanely ambitious, and I finished it! It was basically a buddy comedy where one of the buddies was a hologram. 150 effects shots. WHAT WAS I THINKING?

What are you learning/improving right now?

I'm working on learning more about rigging and game creation.

What project have you seen that you wish you worked on?

I was pretty blown away by Two Face in the Batman series. That looks like some hardcore matchmove! I'd also like to work on anything at all at Imaginary Forces. I love their work.

3D SCENES

One area that I had difficulty with when I first moved into 3D from 2D is how you create a complete scene. Not so much because of having any kind of difficulty filling an area, but more of how you *stop* filling an area. 3D is a complete version of a world, so where would it end? This chapter will be addressing that; we'll take a look at landscapes, backgrounds, and how you control a scene before it gets out of control.

Also in this chapter we'll begin exploring the path from Cinema 4D (C4D) back to After Effects to finalize projects. We'll look at rendering projects from Cinema 4D, preparing C4D projects for After Effects, and the various ways the two software packages talk to each other.

Units

When working on a scene in 3D, there's often a good deal of value in appropriately scaling the scene to make it feel like the real world. So one of your first visits in C4D should be to the *Preferences>Units* to establish the correct measurement form for you. Since I am from the United States, I'll use feet or inches.

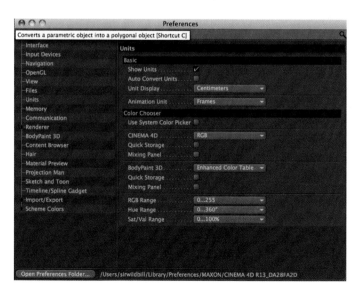

Figure 11.1 Establish the Units in the Preferences window.

For some purposes pixels are a good strategy, but in general, use the measurement form most comfortable for you. In Cinema 4D versions before R12, the units don't have as much of an impact on your scenes as they do now.

Completing Our Pest Control Logo with After Effects

1. Open the *Pest Control* animation from Chapter 8, but this time we won't be using the background object we created before, so you can delete it. Now we'll prepare to render this scene.

2. Go to *Render Settings>Output* and choose the *Preset* for the video format you want. Under *Frame Range* choose the *From* frame number and *To* frame number that you want.

3. Let's move to the *Save* setting. Name the *File* and choose an *Output Format.* I chose *Quicktime.* Turn on *Alpha Channel.* Open *Compositing Project File* and choose *After Effects* for the *Target Application.* I turned on *Include 3D Data.*

4. From *Effect...* choose *Object Glow* and *Scene Motion Blur* (or if you are using R13, *Sub-Frame Motion Blur*).

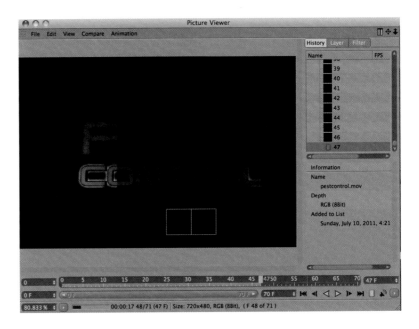

5. Go ahead and choose *Render To Picture Viewer.*

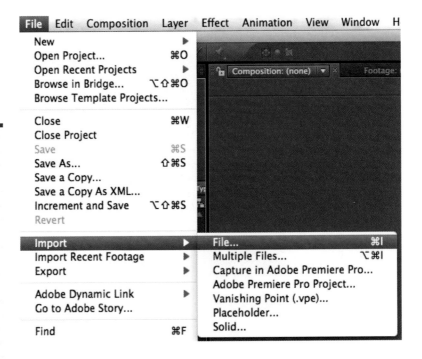

TIP

In order to use .aec files in After Effects you will need to install the Cinema 4D plug-in. The installer can be found inside the MAXON folder in the Exchange Plugins folder. You can always visit the MAXON web site and download the latest version that corresponds to After Effects if necessary. Once you complete the install process, After Effects will need to be restarted.

6. Go to *File>Import* in After Effects and choose the .aec file that Cinema 4D created. This will open a folder with a timeline prepared for Cinema 4D.

7. Once you have imported the .aec file, we can work with our C4D project elements in After Effects. Based on how we rendered, we have two compositions, one for our *Camera* and *Light* layers, one main composition, and then a Quicktime (with an alpha channel) of our content.

8. To enhance our content, I added the *Bevel Alpha* effect, which added a little more depth to our type.

9. To complete this logo animation, I created this microbiology-looking abstract background with solid layers, and the preset *Germs* from the *Effects and Presets* menu under *Animation>Backgrounds>Germs*. I layered them, and added the *Colorama* effect to add another twist.

Creating a Landscape Scene with Integrated Text

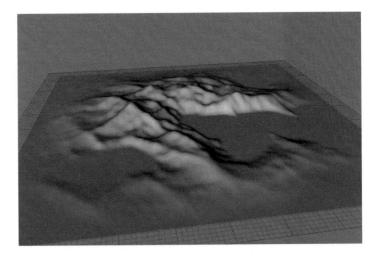

1. Begin by creating a new landscape object. Adjust its size to fit an appropriate area; for example, I made mine about 9000 in width and depth and set the height to be about 1000. I made the sea level about 60%.

2. Next I applied a grassy material to the landscape.

3. Now we'll add a Sky object. Go to *Create>Object>Environment> Sky* to make a sky object. In the *Attribute Manager* for the sky object, you can select a preset for your sky's appearance.

4. We will now start creating our text objects. Go to your *Splines* and choose *Text* from the spline menu. Under *Object* in the *Attribute Manager,* change it to say what you'd like your type to read. Now make the text spline a child of an *Extrude NURBS* object. Make the x and z *Movement* in *Extrude NURBS' Attribute Manager* 120 ft to extrude out the letters. Implant your type somewhere deep into the mountains.

5. I created a green material with a low level of reflection on it and applied it to my text. Then I created a new *Target Light* to shine directly on my text.

6. To add the next word, duplicate the text with Extrude NURBS, along with the light and the target, and move it to a new location. In the *Text Object*'s *Attribute Manager*, I change the text to state what I'd like the second word to read.

7. Repeat what you did in step 6 to create the third word.

8. Create a new *Camera* and have it fly through your scene, stopping periodically to give the audience a chance to read each word.

9. Just like in our previous tutorial, I go to the render settings in Cinema 4D and output an After Effects composition. Except this time we don't need an alpha channel to do what I'm planning here, which is a color treatment and vignette. Once you are ready, press *Render to Picture Viewer*.

10. In After Effects, I added an *Adjustment Layer* with *Hue/ Saturation* set to a lower *Saturation*. Then I added a black solid layer, with an elliptical mask that is inverted, with high value for *Mask Feather*.

Creating a Complete Medical Scene with ZBrush and After Effects

In the following tutorial, we will quickly create a medical scene simulating red bloods cells. So let's begin in ZBrush.

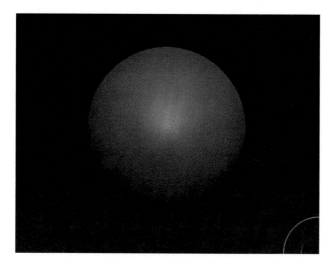

1. In ZBrush, create a new sphere. Click *Make Polymesh 3D* and press the letter T to use the *Edit* mode.

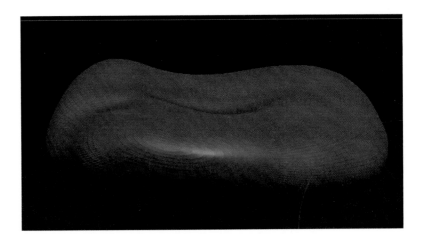

2. Choose the *Move* brush, and set your *Draw Size* to about 160. Push the sphere down until it is still a circular, but flattened, irregular shape.

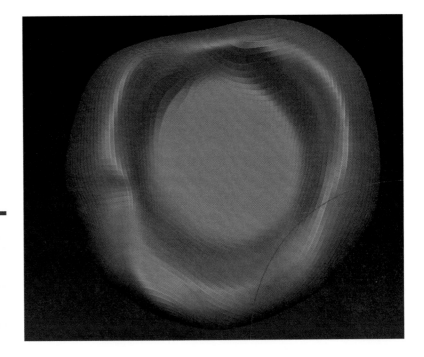

TIP

When working in ZBrush, you may feel the need to have more polygons to work with; you can always add more subdivisions using GeometryHD>DivideHD. Once you've done this, you can get a higher resolution preview of your work by clicking Sculpt HD.

3. Switch to ZSub, which is a subtractive mode. Select the *Mallet* brush, and use a high level of *Intensity* and *Draw Size*. Create a circular indent like that shown in the previous image.

4. We are ready to export from ZBrush. Go to the *Tool* menu, and press *Export*. This will make an .obj file. My plan is to apply materials in Cinema 4D so I can import that file into Cinema 4D. If you have the GoZ plug-in, you can use a single click to hop over to Cinema 4D. The GoZ plug-in is a free download from Pixologic's web site.

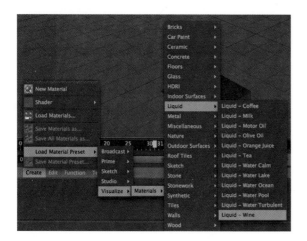

5. With our cell object in Cinema 4D, I find the *Visualize>Materials> Liquid>Liquid–Wine* gets us very close to a blood look.

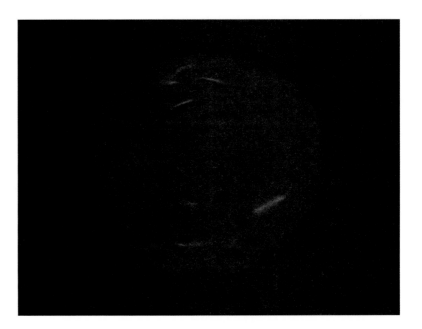

6. To get it where I want it to be, in the *Materials Manager*, I'll make the *Color* a brighter red with a *Brightness* setting of around 200%. Under *Diffusion* I increased the *Mix Strength* to around 80%. Under *Bump* I brought the *Strength* to about 130%.

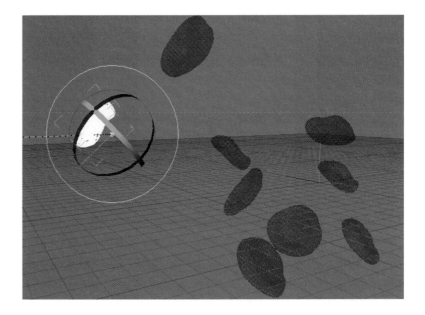

7. Make several duplicates, and animate the *Position* and *Rotation* parameters for each.

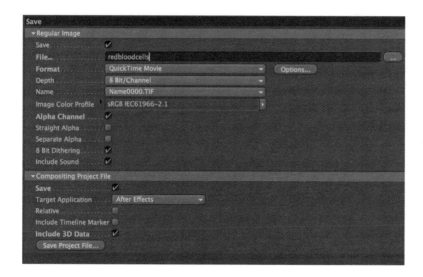

8. As we have been doing all throughout this chapter, let's render our timeline. Be sure to turn on *Alpha Channel* and turn on *Save* for *After Effects*.

9. I made a cloudy red background in Photoshop and I composited the red blood cells over the painting. I also added some color correction.

COMPOSITING

In Chapter 11 we explored how After Effects and Cinema 4D (C4D) can be combined to create complete scenes; in this chapter we'll go a little deeper into that subject by creating more complex projects. We'll be exploring Multi-Pass rendering and really take advantage of the special relationship between Cinema 4D and After Effects.

Creating a 3D Video Wall Using the Object Buffer

1. In Cinema 4D I've modeled a screen, mounted like a drive-in, in the middle of nowhere. I know it doesn't quite make sense, but it will illustrate this point.

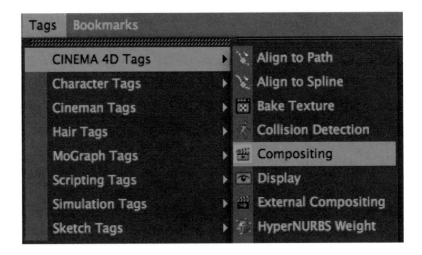

2. Select the screen object from the *Object Manager* and go to *Tags>Cinema 4D Tags>Compositing*. By applying the *Compositing* tag we've identified this object as one for which we want to create a separate pass, and treat it separately from the rest of the scene.

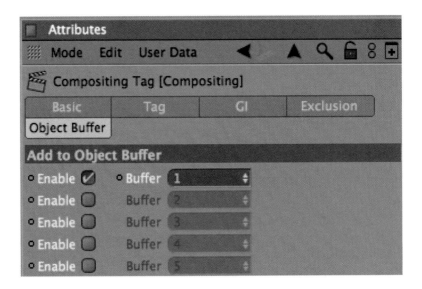

3. In the *Attribute Manager* for the *Compositing Tag*, select *Object Buffer*. Turn on *Enable*. The first buffer will be activated for *Buffer 1*.

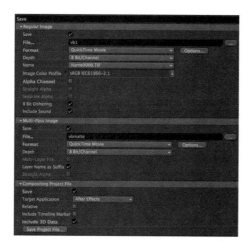

4. Let's go to the *Render Settings.* Under *Regular Image,* turn on *Save.* Give the *File* a name. Set the *Format* to *Quicktime Movie.* Turn on *Multi-Pass Image* and set the format to *Quicktime Movie* (for this example I want to match the format settings I am using at the top for *Regular Image*); give the second pass its own filename. Finally, turn on *Save* under *Compositing Project File* and set *After Effects* as the *Target Application.* Turn on *Include 3D Data.*

5. Click the *Multi-Pass* button. Here's where we can set up different passes for all kinds of different parameters of C4D. Choose *Object Buffer.* Under *Group ID* we need to connect the buffer to 1. Every object where we set *Buffer 1* will now be included in this pass. In this case it is just the single object, the screen. Now we can go ahead and *Render to Picture Viewer.*

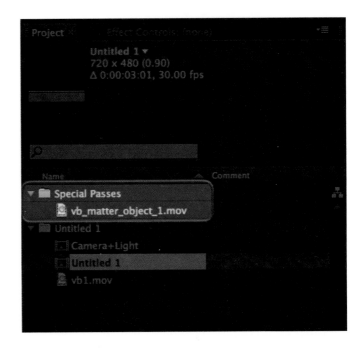

6. Import the .aec into After Effects. Notice the two folders; *Special Passes* will have our compositing matte.

7. I removed the precomp with *Camera+Light* since we won't need them for this project. I imported the itwillburntitle.mov and placed it in the timeline over the regular image from C4D. On the top I drag and drop the matte pass from C4D.

8. Switch to the *Modes* and on the itwillburntitle.mov layer, go to *Track Matte* and choose *Luma Matte 'vb_matter_object_1.mov'*.

9. So now we have our matte applied to the footage, but the angle of the screen looks wrong for the shape of the footage since we are only cutting out the video but not adjusting its place in 3D space. Apply *Effects>Distort>Corner Pin* to the itwillburntitle.mov. Set four corners of the footage and there you go—now our video is well integrated with our 3D scene.

Getting Great Depth of Field Using Multi-Pass Rendering and After Effects' Lens Blur

1. For this tutorial we'll return to the text landscape project from Chapter 11. However we will use different render settings. To get started I put the *Compositing* tag on the *Sky* object and set *Object Buffer* to 1.

2. In *Render Settings* I've set up my render for *Multi-Pass* and an After Effect *Compositing File.*

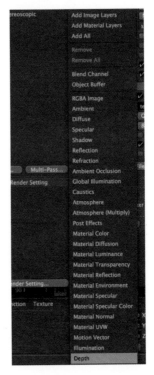

3. Click the *Multi-Pass* button and add *Depth.*

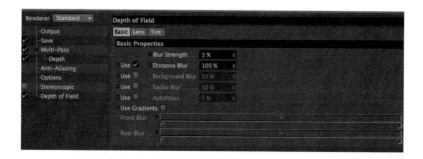

4. Click *Effect* and add in *Depth of Field*. Now we can *Render To Picture Viewer*.

5. Import the .aec into After Effects. Create a timeline with the regular pass and depth pass over it.

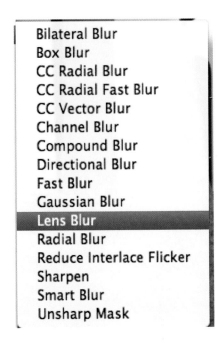

6. Apply *Effects>Blur and Sharpen>Lens Blur* to the regular pass layer.

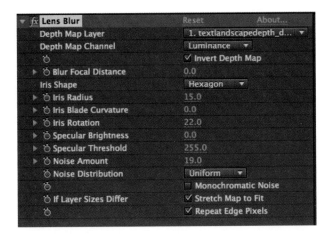

7. Turn off the visibility of the depth pass layer. In the *Lens Blur* effect, set the *Depth Map Layer* to be the depth pass layer. Turn on *Invert Depth Matte.* Add a little *Noise Amount* for a nice film-like effect. Also, be sure to turn on *Repeat Edge Pixels* to avoid an ugly effect along the edge.

8. Animate the *Iris Radius* over the course of the timeline to keep what you want in focus at that point in time. This should look pretty cool.

Cinema 4D's special relationship with After Effects goes even further than the areas we've covered so far. Next, we'll take a look at the *External Compositing* tag. Like the *Compositing* tag, this tag allows you to render data that is specific to compositing in objects later using After Effects. But this does even better—rather than just give you a *Track Matte*, this will create a *Solid* layer that can be replaced with content. The *Solid* layer will have *Position* and *Rotation* animation information.

Deeper Compositing Control Using the External Compositing Tag

1. For this, I'll return to our video board project. I have a camera animated to push in toward the screen and then turn after it lands.

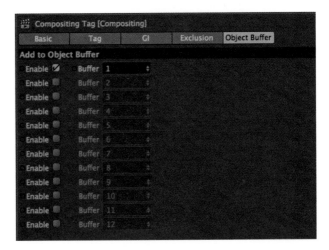

2. Just like in our first tutorial with this scene, I am adding the *Compositing* tag to the screen object, and assigning it *Object Buffer>Buffer 1*.

3. In the *Object Manager* add the *External Compositing* tag to our screen object. This tag can be found in *Tags>Cinema 4D Tags>External Compositing*.

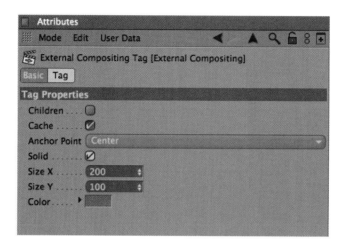

4. In the *External Compositing* tag's *Attribute Manager,* turn on *Solid.* Match the *Size X and Y* to your object. This is pretty easy to do; with the screen object highlighted, look at the numbers for size in the *Coordinate Manager.*

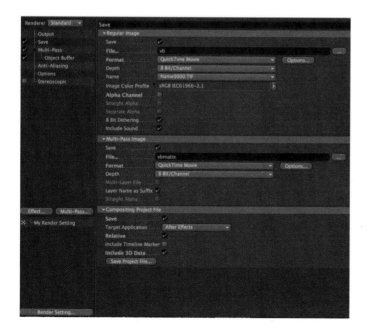

5. Go to the *Render Settings* and set up your render in the same way we have done previously for an After Effects composition with *Object Buffer* for *Group ID 1.* Then use *Render To Picture Viewer.*

6. Import the .aec file into After Effects. Open up the main composition; inside find the precomp for *Camera+Light* and in there you'll find the *Screen* solid. Import the itwillburntitle.mov.

7. Highlight the solid screen layer. Hold down Option, and drag the itwillburntitle.mov to the screen solid. It will now replace the solid with our footage. Use the *Scale* tool to adjust the footage so it will fit appropriately.

8. Step back out to the main composition, and you may see that it still is not consistently keeping its size relative to the actual 3D model of the screen. Bring the matte pass from the *Special Passes* folder into the timeline and place it above the *Camera+Light* precomp. Make it a *Track Matte* for the *Camera+Lights* precomp layer, and set it to *Luma Matte*. This will keep it trim to the edge.

9. If you'd like to get some of the edge of the screen back into the scene, reduce the size of the matte layer by about 2% of *Scale*. Then to soften the edge a little bit, I added *Blur & Sharpen>Fast Blur* with *Blurriness* set to 7.

Final Thoughts

The decision for most motion graphics designers to jump over into the realm of 3D can be intimidating. However, the purpose of this book is to bring users over that divide gradually. First, when getting to this point, remember that you don't have to learn every detail of a piece of software. Rather you should gradually add techniques to your skillset. It's okay to learn just a few things and add more over time. Second, I highly recommend maxing out software like After Effects before getting into true 3D. Why? Because software like After Effects is very versatile, and it may hold within it the ability to do the things you need it to without you having to add another expensive software package with a learning curve. If you become overwhelmed and stressed by this, you won't make the most use of what these great tools have to offer.

Additionally I highly recommend starting your 3D tool kit with Maxon's Cinema 4D because it has the kindest learning curve I've seen on a 3D software package. Its tools will make sense to the advanced After Effects user, and the areas that aren't immediately intuitive are still easier to pick up than other 3D packages. I very much hope that you have enjoyed this book, and I hope that you continue your adventure into the third dimension beyond the scope of the introductory projects highlighted in this text.

INDEX

Note: Page numbers followed by *f* indicates figure and *t* indicates table.